Praise for David Weinberger's *Too Big to Know*

"David Weinberger, one of the earliest and most perceptive analysts of the Internet, thinks we are looking at the wrong thing. It is not the content itself, but the structure of the Internet, that is the important thing. At least, as far as the destruction of a millennia-long human project is concerned. . . . Mr. Weinberger's book will likely make a splash and be widely discussed, as it deserves to be." **—Quentin Hardy, Deputy Technology Editor,** *New York Times*

"If anyone knows anything about the web, where it's been and where it's going, it's David Weinberger. . . . *Too Big to Know* is an optimistic, if not somewhat cautionary tale, of the information explosion that is just underway. Weinberger says that networked knowledge may bring us closer to the truth, but it certainly brings us closer to the truth about knowledge. In the end the changing shape of the container of ideas, from scarcity to abundance, is likely to change far more than we can ever imagine." **—Steven Rosenbaum,** *Forbes*

"Sober, illuminating, and wry." *—Guardian*

"In *Too Big to Know: Rethinking Knowledge Now that the Facts Aren't the Facts, Experts Are Everywhere, and the Smartest Person in the Room Is the Room*, the simultaneously fascinating and frustrating book by Berkman Center senior researcher David Weinberger, there is a wonderful moment where the mechanisms of 'factbuilding' are laid bare. . . . *Too Big to Know* is a smart, readable book." **—C. W. Anderson,** *The Atlantic*

"Insightful. . . . Weinberger, a senior researcher at Harvard's Berkman Center for Internet and Society, doesn't offer tidy solutions to the dilemmas posed by easy access to near-infinite information. Instead, he's written a guidebook to this brave new universe of knowledge." **—Hiawatha Bray,** *Boston Globe*

"Weinberger's intention [seems to have been] to shock his audience into realizing that the rules of the ontological game may be changing." *—The Observer*

"*Too Big to Know* is an entertaining guide to the new world of networked knowledge, but it's also a useful reminder of the sheer diversity of network types necessary to meet the challenge that the smartest person in the room be the room itself." **—Mark Jones,** *Reuters News* **journalist**

"A long, fascinating account of how knowledge itself changes in the age of the Internet—what it means to know something when there are millions and billions of 'things' at your fingertips, when everyone who might disagree with you can find and rebut your assertions, and when the ability to be heard isn't tightly bound to your credentials or public reputation for expertise."

—Cory Doctorow, *Boing Boing*

"Will this be one of the big books of 2012? Probably."

—Tyler Cowen, *Marginal Revolution*

"[W]hat Weinberger has done in a relatively short, admirably readable book is take on a complex, shifting, destabilized, and boundary-free set of ideas and talk about them in a way that makes them connect in a way that gives them coherent shape. He has used the edges provided by the format of the book to organize a lot of material into a series of linked observations, even as he prompts us to think 'yes, but' and 'of course, there's also . . .' and in the final analysis encourages us to think about the way knowledge behaves today and where it may lead us."

—*Inside Higher Ed*

"[Weinberger] eschews the pedantic tone of similar books for a studied reflection on the Web, delivered with a wink. . . . Weinberger's studies provide clearly lettered signposts in the directions future studies would do well to aim themselves."

—*Shelf Awareness*

"Well written, with pithy thoughts, and not over-demanding. . . . The one truly important point, and the one this book ends on, is that access to knowledge is distinct from creating knowledge. No technology can substitute for the slow and painstaking work of Charles Darwin before he published *On the Origin of Species by Means of Natural Selection, or, the Preservation of Favoured Races in the Struggle for Life*. No amount of linking substitutes for investigative journalism. Assuming otherwise is the real danger the ability to access knowledge online poses for us."

—*The Enlightened Economist*

"This is an ambitious thesis. . . . [A] book about the prospect of yet another digital revolution."

—*The Daily*

"[A] seminal new book."

—*KM World*

"A breathtaking review of what it means to be in a world that's powered by connection." —Seth Godin, via Squidoo

"*Too Big to Know* reminds us of the incredible richness of the Internet and shows how the networked contributions of many experts combine to create a new kind of knowledge." —Center for Information-Development Management

"Weinberger's work carries a learned gravitas that others rarely approach." —5000 Blog, *Brian J. McNely*

"The book is excellent in the sense that it encourages us to think deeply about the messy nature of epistemology—yes, that's an opinion and not a fact!" —Rationally Speaking Blog, Greg Linster

"*Too Big to Know* is an important book and has major implications for medical education. We should talk and work to figure out how stopping points and our shifting understanding of knowledge applies to the next generation of doctors. Because for the first time in history, there's just too much for a doctor to know." —33Charts.com, Bryan Vartabedian, MD

"Razor-sharp analysis of the state of knowledge in the age of computer networking. . . . A witty and wise companion in this new age of information overload." —*Kirkus Reviews*

"[Weinberger] engagingly examines the production, dissemination, and accessibility of knowledge in the Internet era. . . . Weinberger's book is full of relevant and thought-provoking insights that make it a must-read for anyone concerned with knowledge in the digital age." —*Publishers Weekly*

"[I]nsightful. . . . A thought-provoking work that is both reassuring and daunting, this will appeal to all readers but will be of special interest to anyone studying information technology. Recommended." —*Library Journal*

"David [Weinberger]'s new book is a wonderfully unsettling piece—it challenges our notion of what knowledge is, and introduces the uncomfortable question of how we navigate this new space. . . . [T]his may be the most unsettling and radical book David's put forth." —Ethan Zuckerman, Director of the MIT Center for Civic Media

"David is an intellectual hero of mine. . . . He is open and curious. He does this with charm and unwarranted but sincere self-deprecation. All that comes across in his books. . . . These are profoundly disruptive ideas about ideas. It helps that they come from someone who presents them via doubt rather than dogma. David is, like me, essentially an optimist, but he sees the choices we have and the dangers that present themselves if we chose the wrong paths."

—Jeff Jarvis, author of *Private Parts* and *What Would Google Do?*

"*Too Big to Know* is Weinberger's brilliant synthesis of myriad little debates—information overload, echo chambers, the wisdom of crowds—into a single vision of life and work in an era of networked knowledge."

—Clay Shirky, author of *Here Comes Everybody* and *Cognitive Surplus*

"*Too Big to Know* is a refreshing antidote to the doomsday literature of information overload. Acknowledging the important roles that smart mobs and wise crowds have played, David Weinberger focuses on solutions to the crisis in knowledge—translating information into new knowledge by exploiting the network. Based upon the premise that 'knowledge lives not in books, not in heads, but on the net,' Weinberger outlines a bold net infrastructure strategy that is inclusive rather than exclusive, creates more useful information—metadata, exploits linking technologies, and encourages institutional participation. The result is a network that is both 'a commons and a wilds' where the excitement lies in the limitless possibilities that connected human beings can realize."

—David S. Ferriero, Archivist of the United States

"David Weinberger's *Too Big to Know* is an inspiring read—especially for networked leaders who already believe that the knowledge to change the world is living and active, personal, and vastly interconnected. If, as David writes, 'Knowledge is becoming inextricable from—literally unthinkable without—the network that enables it' our great task as leaders is to design networks for the greater good. David casts the vision and gives us excellent examples of what that looks like in action, even as he warns us of the pitfalls that await us."

—Tony Burgess, Cofounder, CompanyCommand.com

"*Too Big to Know* is a stunning and profound book on how our concept of knowledge is changing in the age of the Net. It honors the traditional social practices of knowing, where genres stay fixed, and provides a graceful way of understanding new strategies for knowing in today's rapidly evolving, networked world. I couldn't put this book down. It is a true tour de force written in a delightful way."

—John Seely Brown, co-author of
The Social Life of Information and *A New Culture of Learning*

"With this insightful book, David Weinberger cements his status as one of the most important thinkers of the digital age. If you want to understand what it means to live in a world awash in information, *Too Big to Know* is the guide you've been looking for."

—Daniel H. Pink, author of *Drive* and *A Whole New Mind*

"Led by the Internet, knowledge is now social, mobile, and open. Weinberger shows how to unlock the benefits."

—Marc Benioff, chairman, CEO salesforce.com,
bestselling author of *Behind the Cloud*

TOO BIG TO KNOW

TOO BIG
TO KNOW

*Rethinking Knowledge
Now That the Facts Aren't the Facts,
Experts Are Everywhere,
and the Smartest Person in the Room
Is the Room*

David
Weinberger

BASIC BOOKS
*A Member of the Perseus Books Group
New York*

Hardcover first published in 2011 by Basic Books,
A Member of the Perseus Books Group
250 West 57th Street, 15th Floor
New York, NY 10107
Paperback first published in 2014 by Basic Books

Books published by Basic Books are available at special discounts for bulk purchases in the United
States by corporations, institutions, and other organizations. For more information, please contact
the Special Markets Department at the Perseus Books Group, 2300 Chestnut Street, Suite 200,
Philadelphia, PA 19103, or call (800) 255-1514, or e-mail special.markets@perseusbooks.com.

The Library of Congress has cataloged the hardcover as follows:
Weinberger, David, 1950–
    Too big to know : rethinking knowledge now that the facts aren't the facts, experts are
everywhere, and the smartest person in the room is the room / David Weinberger.
        p. cm.
    Includes bibliographical references and index.
    ISBN 978-0-465-02142-0 (alk. paper)—ISBN 978-0-465-02813-9 (e-book) 1. Information
technology—Social aspects. 2. Internet—Social aspects. 3. Knowledge, Sociology of. I. Title.

HM851.W445 2012
303.48'33—dc23

ISBN 978-0-465-08596-5 (paperback)

                                                                                        2011034727

10 9 8 7 6 5 4 3 2 1

CONTENTS

PROLOGUE

The Crisis of Knowledge

THREE OF THE SIX HEADLINES on the front page of the *New York Times* on the day I happen to be writing this (June 21, 2010) could have as their subhead "Knowledge in Crisis!"

The lead story on this randomly chosen day at The Paper of Record takes a long look at the reasons behind the failure of the supposed fail-safe mechanism on the British Petroleum oil rig that fouled the Gulf of Mexico.[1] The five authors explain clearly what a "blind shear ram" is ("two tough blades . . . poised to slice through the drill pipe, seal the well and save the day"), how close it came to working, and what exactly went wrong. The article bounces us from vivid descriptions of the moments when the equipment proved inadequate, to an extensive examination of the claims made by the oil industry, to a discussion of the internal processes of a lax regulatory agency. The article's controversial upshot: Fail-safe mechanisms sound reassuring but in actuality create a terrifyingly risky "single point of failure."

The subject of the article may be the BP oil spill, but its real topic is the limits of expert knowledge in tackling complex problems. It attempts to explain what we could have done to prevent the disaster, given how hard it is for us to know what will work. How wide is the inevitable gap between our perfect theories and their mechanical imperfection? How much of what we know depends on what we would like to believe? What are the institutional biases that prevent us from acting on what we know? Can the forces that corrupt knowledge be countered, or should we recognize that knowledge is always going to be degraded by politics and greed?

Below this front-page article is a preliminary investigation of John Updike's archive, suggesting that the author researched the settings of his novels with a great concern for accuracy—down to the sales figures for Toyota franchises and the look of Florida license plates. The archive reveals rich details about the private life of a writer who carefully controlled the public view of himself—"a one-man gated community," the article says—but who seems to have prepared for an open house once he died, preserving correspondence and even recording the grades he got on quizzes as an undergraduate at Harvard. The Updike revealed by the archives is at some variance with the Updike we thought we knew. For example, though he encouraged a reputation for revising little, the archives he carefully preserved show that he meticulously reworked his manuscripts.[2]

The story is about John Updike, but it raises the important question of how we are going to understand artists once they no longer leave paper trails. It's the paper that Updike collected and preserved that lets us see how much of his fiction is indebted to researching the facts that populate his characters' world. And it's the paper with Updike's penciled markings that allows us to see the effort behind his effortless prose. What will we be able to know about writers whose drafts and markings vanish in a flow of insubstantial bits? Without such a record, how will we be able to observe, as the *Times* article does, that Updike's early letters pay scant attention to the Korean War and McCarthyism? How will we know that which can be learned only by observing what is *not* mentioned if personal archives become as fragile as the magnetic traces on an aging hard drive?

At the bottom of the front page of this particular edition of the *Times* is a feature about soccer players in the World Cup who fake injury to draw a free kick.[3] The article notes that it would be much easier to nab these "actors" if the referees had access to video replays, but that would require sacrificing the flow and spontaneity of the sport—a political and cultural change that the International Federation of Association Football is reluctant to make.

We could read this as a sports story, of course, but it is also about the complex role that knowledge plays in our world. How much does accuracy matter? How much are we willing to let experts intrude in order to get a better ruling? What are the positive aspects of the fallibility of human knowledge? Do we want to let experts swarm onto every literal and figurative field? Doesn't expertise come with a cost? Does there turn out to be a benefit to letting events have blurry edges of ignorance?

These three newspaper articles from different realms of life are part of a long argument we've been having about knowledge over the roughly 2,500 years since we decided it would be useful to distinguish reliable ideas from mere opinions. Despite the constant disputes, the basics of our system of knowledge are quite well defined. Especially for those who have grown up digital, here's a reminder of how it works:

People study hard and become experts in particular areas. They earn credentials—degrees, publications, the occasional Nobel Prize—that make it easier for us trust them. They write books, teach classes, and go on TV, so that we all can benefit from their hard work. The results of that work go through vetting processes appropriate to the type and importance of its claims, providing us with even more assurance of its accuracy. As new discoveries are made and sanctioned, the body of knowledge grows. We build on it, engaging in a multi-generational project that, albeit with occasional missteps, leads us further along in our understanding of the world. Knowledge is a treasure, knowing is the distinctively human activity, and our system of knowledge is the basis for the hope that we might all one day come to agreement and live in peace.

We've grown up thinking that this is how knowledge works. But as the digital age is revealing, that's how knowledge worked when its medium was paper. Transform the medium by which we develop, preserve, and communicate knowledge, and we transform knowledge.

Within those three front-page *New York Times* stories we can already discern challenges to some of our most basic ideas about what knowledge is and how it works:

In the three months before it was finally capped, BP's gushing oil could be seen live on any Web site that cared to embed the video, surrounded by whatever text and links the page's owner thought important to understand it. The online version of the *New York Times* article linked to its source data, including "previously unreleased notes scrawled by industry crisis managers,"[4] in case we wanted to become our own experts over the course of breakfast. Every blogger is a broadcaster, and every reader is an editor.

Updike's archive "may be the last great paper trail," as the article says, and every reader who uses a word processor—or, as most of us say these days, who writes—contrasts the solid traces we used to leave behind with the digital dust we currently leave in our wake: more of it for sure, but also more likely to be blown away by a hard-drive failure or a change from floppies to CDs to DVDs to Blu-Rays to whatever comes next and next after that. A paper archive like Updike's seems so quaint these days—so manageable in size, so under the control of the person who is its subject. What are people going to know about us when they are left to rake through the acres of drafts and photos we've strewn across our hard drives and Facebook pages?

Every spectator of the World Cup can see the replays that the referees cannot, making the lively online discussion among soccer fans more fact-based and knowledgeable than the decisions of the expert judges on the field.

The crisis in knowledge goes far beyond questions raised by articles on the front page of one morning's newspaper, and far beyond the blurring of lines between readers and editors, authors and biographers, spectators and referees. Our most important institutions are being shaken by questions about knowledge that we thought were as firmly settled as those institutions' marble and concrete foundations:

> Universities are debating whether professors ought to be required to post all their research freely on the Web, rather than (or in addition to) publishing it in prestigious but expensive journals. Further, should a professor who is shaping the discipline's discussion through her mighty participation in online

and social media get tenure even if she hasn't published sufficiently in peer-reviewed journals?

Librarians are enmeshed in a struggle for a workable vision of a future for their institutions, not only debating the merits of new techniques for navigating collections but wondering how to weigh the expertise of the "crowd" against that of those with credentials.

Major business consulting firms—once charged with preparing shiny, conclusive reports—now experiment with providing clients with access to a network of experts who represent a divergent range of opinions.

Business leaders looking at the overwhelming amount to know in a globalized world are trying new decentralized decision-making processes that more effectively take advantage of the expertise spread out across their networks—modeling themselves on the distributed leadership becoming common at large, Web-based collaborative projects such as Wikipedia.

The US intelligence agencies and the State Department are caught up in internal battles between the old "need to know" culture and the new "need to share" mindset. The executive branch of the US government is struggling to define exactly how much and what type of information its agencies should release to its citizens.

The sciences find themselves both being enhanced by the efforts of amateurs and having to defend their reliability against amateurs—sometimes fiercely partisan ones—who have access to the same data as the professionals. Even among many respected scientists, the traditional journals are beginning to look like blockages in the system of knowledge, for they can publish so few of the worthwhile submissions. The journal *Nature*—at the top of the prestige pyramid—has begun its own online site where it publishes without regard to page count, to compete with the new generation of open access journals that have rapidly grown in stature and importance.

As for the media, you can hardly get them to stop talking about what they're going to do about the Internet where there are no editors, and where the old media are perceived as biased and self-involved.

At its worst, this crisis of knowledge is apparent in the jumble of fears put forward as obvious: The Internet is an unedited mash of rumor, gossips, and lies. It splinters our attention and spells the end of reflective, long-form thought. Our children don't read any more. They certainly don't read newspapers. Everyone with any stupid idea has a megaphone as big as that of educated, trained people. We form "echo chambers" online and actually encounter fewer challenges to our thinking than we did during the broadcast era. Google is degrading our memories. Google is making us stupid. The Internet loves fervid, cult-driven amateurs and drives professionals out of business. The Internet represents the ascension of yahoos, a victory lap for plagiarists, the end of culture, the beginning of a dark ages inhabited by glassy-eyed chronic masturbators who judge truth by the number of thumbs up, wisdom by the number of views, and knowledge by whatever is the most fun to believe.

And yet, at the very same time, sites such as Politifact.com are fact-checking the news media more closely and publicly than ever before, and Jodi Kantor, a reporter for the *New York Times,* says that knowing that bloggers will go over every word she writes has made her better at her job.[5] Libraries are breaking new ground in using all available data—including contributions from readers—to make it far easier than ever for readers to find and understand the resources they need. Science is advancing at an unheard-of pace thanks to new collaborative techniques and new ways to publish vast amounts of data and troll it for patterns and inferences. Businesses are managing in an era that defies predictions by finding expertise in every corner of their organizations, and across the broad swath of their stakeholders.

So, we are in a crisis of knowledge at the same time that we are in an epochal exaltation of knowledge. We fear for the institutions on which we have relied for trustworthy knowledge, but there's also a joy we can feel pulsing through our culture. It comes from a different place. It comes from *the networking of knowledge.* Knowledge now lives not just in libraries and museums and academic journals. It lives not just in the skulls of individuals. Our skulls and our institutions are simply not big enough to contain knowledge. Knowledge is now a property of the network, and the network embraces businesses, governments, media, museums, curated collections, and minds in communication.

That knowledge is a property of the network means more than that crowds can have a type of wisdom in certain circumstances. And, as we will see, it's not simply that under some circumstances groups are smarter than their smartest member. Rather, the change in the infrastructure of knowledge is altering knowledge's shape and nature. As knowledge becomes networked, the smartest person in the room isn't the person standing at the front lecturing us, and isn't the collective wisdom of those in the room. The smartest person in the room is the room itself: the network that joins the people and ideas in the room, and connects to those outside of it. It's not that the network is becoming a conscious super-brain. Rather, knowledge is becoming inextricable from—literally unthinkable without—the network that enables it. Our task is to learn how to build smart rooms—that is, how to build networks that make us smarter, especially since, when done badly, networks can make us distressingly stupider.

The new way of knowing is just now becoming apparent. Although we can't yet know its adult form, some aspects are taking form. Networked knowledge is less certain but more human. Less settled but more transparent. Less reliable but more inclusive. Less consistent but far richer. It feels more natural because the old ideals of knowledge were never realistic, although it's taken the networking of our culture to get us to admit this.

This book will follow one particular pathway through an impossibly large territory. That's appropriate, since at the core of knowledge's new exaltation and transformation is a straightforward acknowledgment of one basic truth we've always known but that our paper-based system of knowledge simply could not accommodate: The world is far, far too big to know.

TOO BIG TO KNOW

1

Knowledge Overload

Triangular Knowledge

In his 1988 presidential address to the International Society for General Systems Research, Russell Ackoff, a leading organizational theorist, sketched a pyramid that has probably been redrawn on a white board somewhere in the world every hour since.[1] The largest layer at the bottom of the triangle represents data, followed by successively narrower layers of information, knowledge, understanding, and wisdom. The drawing makes perfect visual sense: There's obviously plenty of data in the world, but not a lot of wisdom. Starting from mere ones and zeroes, up through what they stand for, what they mean, what sense they make, and what insight they provide, each layer acquires value from the one below it.

Ackoff was not the first person to propose a data-information-knowledge-wisdom (DIKW) hierarchy. Milan Zeleny had discussed a similar idea in an article published the year before, and Michael Cooley had come up with roughly the same concept in an article written shortly before that. In fact, in 1982, Harlan Cleveland not only described the hierarchy in an article in *The Futurist,* he pointed to its earliest known version:[2]

> *Where is the Life we have lost in living?*
> *Where is the wisdom we have lost in knowledge?*
> *Where is the knowledge we have lost in information?*

T. S. Eliot wrote these lines in 1934 in a poem called "The Rock." The next reference, preceding all of the business articles and books on this topic, appeared in 1979, in a song called "Packard Goose" by Frank Zappa.[3]

Now, no one believes any of these thinkers actually plagiarized T. S. Eliot, much less Frank Zappa. The idea seemed to have a certain inevitability. Imagine you're in charge of your company's data processing center in 1955—well after Eliot's poem and well before Zappa's lyrics—and you're watching the experts from IBM install the most popular corporate computer of the 1950s, the IBM 650.[4] With only seventy-five of these machines installed anywhere, you're on the leading edge. The 650 has the latest in punchcard-reading technology and can calculate 78,000 additions or subtractions per minute. Your 2011 home PC probably does about 300 billion per minute, but back then the 650's computing power didn't come cheap: The 650 cost your company $500,000, equivalent to $4 million in 2011 dollars.[5] It's got its own room, its own fleet of maintenance folks, its own dress code: white lab coats, please. But it's a workhorse, and you'll get good value from it— processing payrolls, calculating projected sales figures, managing the human resources database.

You've gotten used to a particular drill when corporate executives visit your data center and marvel at the boxes of thousands, hundreds of thousands, millions of punchcards. Ah, you patiently explain, that's just data. By itself it has no value. But process the data and you get *information*. Information is to data what wine is to a vineyard: the delicious extract and distillate. In 1955, information was the value of the seemingly senseless mounds of data that were quickly accumulating.

Thirty years go by. You and the rest of the world have been refining data into information by the boxcar-full. Now you are as overloaded with information as you once were with data, and you have the same question: You've spent a lot of money gathering it, but what's its value? Information has become a problem, not a solution. So, how do you justify your investment in producing all that information out of all that data? The same way you justified your investment in data. You've refined the data to produce information, and you've refined the information to generate something of greater value. You've got to call it something. How about "knowledge"? Thus did the knowledge management industry take off in the early and mid-1990s, based on the promise that it would help enterprises discern and share the highest-value information it was generating.

Of course, to get knowledge to look like it's an outcome of information, you have to radically redefine knowledge. For Ackoff, knowledge was know-how that transforms "information into instructions,"[6] such as "knowing how a system works or how to make it work in a desired way."[7] Skip Walter, who was mentored by Ackoff, said that while information is structured data, knowledge "is actionable information." For example, information becomes knowledge when you decide whether to wear a sweater.[8] Milan Zeleny, who beat Ackoff to the punch by a couple of years, said knowledge is like the recipe that turns information into bread, while data are like the atoms that make up the flour and the yeast.[9]

Fine. But when T. S. Eliot wrote "Where is the knowledge we have lost in information?" he was not thinking of knowledge as "actionable information." The knowledge discovered by scientists and researchers isn't a recipe. In fact, it wasn't even information in the current sense of a mass of unrelated facts. Back before Ackoff's pyramid, back when the idea of knowledge first occurred to us, the ability to know our world was the essential difference between us and the other animals. It was our fulfillment as humans, our destiny. Knowledge itself fit together into a perfectly ordered whole. Knowledge therefore was considered for thousands of years in the West to be an object of the most perfect beauty. Indeed, in knowing the world, we were striving to understand God's creation as He himself understands it, given our mortal limitations; to know the world was to read it like a book that God had written explaining how He had put it together. Darwin spent five years sailing on a small boat, Galileo defied a Pope, and Madame Curie handled radioactive materials, all in pursuit of knowledge as the most profound of human goals. That is what knowledge has meant in our culture, and it has little to do with the middle layer of a made-up pyramid that shears knowledge of all but its most prosaic, get-'er-done utility.

Despite this, the DIKW pyramid gets one thing very right about how we've thought about knowledge. Our most basic strategy for understanding a world that far outruns our brain's capacity has been to filter, winnow, and otherwise reduce it to something more manageable. We've managed the fire hose by reducing the flow. We've done this through an elaborate system of editorial filters that have prevented most of what's written from being published, through an elaborate system of curatorial filters that has kept most of what's been published from being shelved in our local libraries and bookstores, and through an elaborate system of professional filters that have kept many of us from being responsible for knowing most of what's made it through the other filters. Knowledge has been about reducing what we need to know.

The Information Age took this strategy and ran with it. We built computers that ran databases that took in a bare minimum of useful information, in predefined categories: first name, last name, social security number, date of birth. . . . Whether they include a dozen fields or a thousand, our information systems worked only because they so rigorously excluded just about everything.

Even the very idea of knowledge began as a way of winnowing claims. In ancient Athens, matters of state were debated in public by any citizen—so long as he was one of the 40,000 free males who had completed military training.[10] Many opinions were expressed—go to war or not, find someone guilty or innocent of a political crime—but only some were worth believing. Philosophers noticed the difference and raised beliefs worth believing into their own class. Plato gave us the abiding formulation: Among all the opinions spouted, the subset that counts as knowledge consists of the ones that not only are true but also are believed for justifiable reasons. That second qualification was necessary because some people hold opinions that are true but only by accident: If you think Socrates is innocent of corrupting the young because you like the way he drapes his toga, your opinion is true but does not constitute knowledge. Hunches and guesses that turn out to be right also aren't knowledge. Knowledge is so important for deciding matters of state—and for understanding who we are and how our world works—that the bar needs to be set high. What makes it over are beliefs we can rely on, beliefs we can build on, beliefs worth preserving and cherishing. That's more or less what Plato, T. S. Eliot, and we today still mean by "knowledge" in its ordinary usage.

We've become the dominant species on our planet because the elaborate filtering systems we've created have worked so well. But we've paid a hidden price: We have raised the bar so high that we have sometimes excluded ideas that were nevertheless worth considering, and false beliefs—once accepted—can be hard to dislodge even after they have been found out. And there is so much that could be known that we just don't have room for it all; our science magazines can publish only so many articles and our libraries can hold only so many books. The real limitation isn't the capacity of our individual brains but that of the media we have used to get past our brains' limitations. Paper-based tools allowed us to write things down, but paper is expensive and bulky. Even once we had computers with millions of times more capacity—your home computer may well have over 300,000 times more memory than the IBM 650[11]—knowledge was hard to transfer from paper, and hard to access on a desktop machine.

Now if you want to know something, you go online. If you want to make what you've learned widely accessible, you go online. Paper will be with us for a long time, but the momentum is clearly with the new, connected digital medium. But this is not merely a shift from displaying rectangles of text on a book page to displaying those rectangles on a screen. It's the connecting of knowledge—the networking—that is changing our oldest, most basic strategy of knowing. Rather than knowing-by-reducing to what fits in a library or a scientific journal, we are now knowing-by-including every draft of every idea in vast, loosely connected webs. And that means knowledge is not the same as it was. Not for science, not for business, not for education, not for government, not for any of us.

Info Overload as a Way of Life

Information overload isn't what it used to be.

Alvin Toffler introduced the idea of information overload to the general public in 1970 in his book *Future Shock*.[12] He positioned information overload as a follow-on to sensory overload:[13] When our environment throws too many sensations at us—say, at a Grateful Dead concert with a light show and the mixed scents of a thousand sticks of incense—our brains can get confused, causing a "blurring of the line between illusion and reality."[14] But what happens when we go up a level from mere sensation, and our poor little brains are buffeted by information?

Toffler pointed to research indicating that too much information can hurt our ability to think. If too many bits of information are transferred into our wetware, we can exceed our "channel capacity"—a term straight out of Information Science. Wrote Toffler: "When the individual is plunged into a fast and irregularly changing situation, or a novelty-loaded context . . . his predictive accuracy plummets. He can no longer make the reasonably correct assessments on which rational behavior is dependent."[15] "Sanity, itself, thus hinges" on avoiding information overload.[16] A term, a fear, and a bestseller were born.

The marketers quickly picked up on the idea, worrying that consumers would get confused if given too much information. But what was too much information? In a study performed in 1974, 192 "housewives" were given information about sixteen different attributes of sixteen different brands. And that information was itself simplified into binary pairs; for example, rather than being told the number of calories, the women were simply told that items were high or low in calories.[17] Yet even this was believed to so overload them with information that they made poor buying decisions. Marketers thus told themselves they were preserving the rational faculties of consumers by strictly limiting how much information vendors provided.

This study strikes us as an artifact of a simpler time. Comparing the calories on sixteen nutritional labels constitutes information overload? We must have been in a very delicate informational state.

The psychological syndrome caused by information overload subsequently got renamed, driven often by nothing more than the desire to market a book. We heard about information anxiety, information fatigue syndrome, analysis paralysis. These debilitating diseases were brought on by data smog, infoglut, and the information tsunami. We were about to drown. Richard Saul Wurman's book *Information Anxiety,* published in 1989, made its case by aggregating startling facts such as "About 1,000 books are published internationally every day"[18] and "Approximately 9,600 different periodicals are published in the United States each year."[19]

We now laugh in the face of such danger. Technorati.com in 2009 tracked over 133 million blogs. Of those, more than 9,600 were *abandoned* every day. With a trillion pages, the Web is far far far larger than anyone predicted. In fact, according to two researchers at University of California–San Diego, Americans consumed about 3.6 zettabytes of information in 2008.[20]

Zettabytes?

This is a number so large that we have to do research just to understand it. Fortunately, these days we have the Internet to answer all our questions, so we can just type "zettabyte" into our favorite search engine and discover that it means 1 sextillion bytes.[21]

Sextillion?

Back to Google. A sextillion is 1,000,000,000,000,000,000,000 bytes. That's 10^{21} bytes. A billion gigabytes times a thousand. Clear?

No? How about this, then: The electronic version of *War and Peace* takes just over 2 megabytes of space on the Kindle. A zettabyte is therefore the equivalent of 5×10^{14} copies of *War and Peace*. Of course, now we have to figure out what 5×10^{14} copies of *War and Peace* look like. Assume each is six inches thick, and they'd stack up to over 47 billion miles. And to understand *that* number, we could point out that it would take light 2.9 days to travel from the front cover of the first volume to the back cover of the last—ignoring the relativistic effects the gravity of this new 250 billion–ton object would create (assuming each volume weighs a pound). Or, put differently, if we divided the novel into two equal parts, *War* would stretch the length of eight trips from the sun to Pluto and *Peace* would stretch eight trips back.

Our little simian brains just can't make sense of numbers like these. But we don't have to have a firm sense of how long a zettabyte would stretch or how much it would weigh, or even how much it could earn just by saving a penny a day, in order to suspect that the change we're seeing in knowledge is not primarily due to the massive increase in the amount of information. Something else is at work.

After all, we've been complaining about what we now call "information overload" for a long time. In 1685, French scholar Adrien Baillet wrote: "We have reason to fear that the multitude of books which grows every day in a prodigious fashion will make the following centuries fall into a state as barbarous as that of the centuries that followed the fall of the Roman Empire."[22] It's comforting to know that the idea that information will cause civilization to come crashing down has survived several crashed civilizations.

Baillet was not some isolated crank. In 1755, none less than the creator of the first modern encyclopedia, Denis Diderot, reasoned: "As long as the centuries continue to unfold, the number of books will grow continually," so "one can predict a time will come when it will be almost as difficult to learn anything from books as from the direct study of the whole universe."[23] And it wasn't just the French who feared they would drown in a sea of leather-bound volumes. In 1680, the German philosopher Gottfried Leibniz wrote of his fear of the "horrible mass of books which keeps on growing"[24] that would someday make it impossible to find anything. This did not prevent him from adding his own dense works to that horrible mass. It never does.

We can trace it back further, if we want. The Roman philosopher Seneca, born in 4 BCE, wrote: "What is the point of having countless books and libraries whose titles the owner could scarcely read through in his whole lifetime? That mass of books burdens the student without instructing."[25] In 1642, Jan Amos Comenius complained that "bookes [sic] are grown so common . . . that even common country people, and women themselves are familiarly acquainted with them."[26]

Of course, those voices sound like whiners to us now. They were drowning facedown in puddles of information. In our age, information overload has blown past every dire prediction, with the zettabyte study from UCSD zooming past the previous estimate of 0.3 zettabytes from a study just two years earlier.[27] The difference between 0.3 and 3.6 zettabytes is ten times the total number of grains of sand on the earth—although these studies probably only chart an unimaginable gap in how they measure information.

It doesn't matter. No matter how long the line of copies of *War and Peace* gets, or how good your intentions, you're still probably not going to get through a single one of them this summer. It doesn't really matter if that bookshelf continues to Pluto and then loops back another fifteen times. An overload of an overload is still just an overload. Does it matter whether you drown in water that's 10 feet deep or stretches for 10^{21} more feet (189 quadrillion miles)?

Yet, something odd has happened. As the amount of information has overloaded the overload, we have not proportionately suffered from information anxiety, information tremors, or information butterflies-in-the-stomach. Information overload has become a different sort of problem. According to Toffler, and for three decades following *Future Shock*'s publication, it was a psychological syndrome experienced by individuals, rendering them confused, irrational, and demotivated. When we talk about information overload these days, however, we're usually not thinking of it as a psychological syndrome but as a cultural condition. And the fear that keeps us awake at night is not that all this information will cause us to have a mental breakdown but that we are not getting enough of the information we need.

So, we have rapidly evolved a set of technologies to help us. They fall into two categories, algorithmic and social, although most of the tools available to us actually combine both. Algorithmic techniques use the vast memories and processing power of computers to manipulate swirling nebulae of data to find answers. The social tools help us find what's interesting by using our friends' choices as guides.

The technologies will continue to advance. This book does not focus on the technological side. Instead, we will pursue a different, and more fundamental, question: How has the new overload affected our basic strategy of knowing-by-reducing?

Filtering to the Front

If we've always had information overload, how have we managed? Internet scholar Clay Shirky says: "It's not information overload. It's filter failure."[28] If we feel that we're overwhelmed with information, that means our filters aren't working. The solution is to fix our filters, and Shirky points us to the sophisticated tools we've developed, especially social filters that rely upon the aggregated judgments of those in our social networks.

Shirky's talk of filter failure intends to draw our attention to the continuity between the old and the new, bringing a calming voice to an overheated discussion: We shouldn't freak out about information overload because we've always been overloaded, in one way or another. But when I asked him about it, Shirky agreed without hesitation that the new filtering techniques are disruptive, especially when it comes to the authority of knowledge. Old knowledge institutions like newspapers, encyclopedias, and textbooks got much of their authority from the fact that they filtered information for the rest of us. If our social networks are our new filters, then authority is shifting from experts in faraway offices to the network of people we know, like, and respect.

While I agree with Shirky, I think there's another—and crucial—difference in the old and new filters.

If you are on the Acquisitions Committee of your town library, you are responsible for choosing the trickle of books to buy from the torrent published each year. Thanks to you, and to the expert sources to whom you look, such as journals that preview forthcoming titles, library patrons don't see the weird cookbooks and the badly written personal reminiscences that didn't make the cut, just as the readers of newspapers don't see the crazy letters to the editor written in crayon. But many of the decisions are harder. There just isn't room for every worthwhile book, even if you had the budget. That's the way traditional physical filters have worked: They separate a pile into two or more piles, each physically distinct.

The new filters of the online world, on the other hand, remove clicks, not content. The chaff that doesn't make it through the digital filter is still the same number of clicks away from you, but what makes it through is now a single click away. For example, when Mary Spiro of the *Baltimore Science News Examiner* posts the "eight podcasts you shouldn't miss,"[29] she brings each of those eight within one click on her blog. But the tens of thousands of science podcasts that didn't make it through her filter are still available on the Net. It may take you a dozen clicks to find Selmer Bringsjord's podcast "Can Cognitive Science Survive Hypercomputation?," which didn't make it through Spiro's filter, but it's still available to you in a way that a manuscript rejected by your librarian or print-book publishers is not. Even if Bringsjord's paper is the millionth result in a Google search, a different search might pop it to the top, and you may well find it through an email from a friend or on someone else's top-ten list.

Filters no longer filter out. They filter *forward,* bringing their results to the front. What doesn't make it through a filter is still visible and available in the background.

Compare that to your local library's strategy. In the United States, 275,232 books were published in 2008, a thirty-fold increase in volume from 1900.[30] But it's highly unlikely that your local library got hundreds of times bigger during those past 110 years to accommodate that growth curve. Instead, your library adopted the only realistic tactic, each year ignoring a higher and higher percentage of the available volumes. The filters your town used kept the enormous growth in book-based knowledge out of sight. As a result, library users' experience of the amount of available knowledge didn't keep up with its actual growth. But on the Net, search engines answer even our simplest questions with more results than the total number of books in our local library. Every link we see now leads to another set of links in a multi-exponential cascade that fans out from wherever we happen to be standing. Google lists over 3 million hits on the phrase "information overload."[31]

There was always too much to know, but now that fact is thrown in our faces at every turn. Now we *know* that there's too much for us to know. And that has consequences.

First, it's unavoidably obvious that our old institutions are not up to the task because the task is just too large: How many people would you have to put on your library's Acquisitions Committee to filter the Web's trillion pages? We need new filtering techniques that don't rely on forcing the ocean of information through one little kitchen strainer. The most successful so far use some form of social filtering, relying upon the explicit or implicit choices our social networks make as a guide to what will be most useful and interesting for us. These range from Facebook's simple "Like" button (or Google's "+1" button) that enables your friends to alert you to items they recommend, to personalized searches performed by Bing based on information about you on Facebook, to Amazon's complex algorithms for recommending books based on how your behavior on its site matches the patterns created by everyone else's behavior.

Second, the abundance revealed to us by our every encounter with the Net tells us that no filter, no matter how social and newfangled, is going to reveal the complete set of knowledge that we need. There's just too much good stuff.

Third, there's also way too much bad stuff. We can now see every idiotic idea put forward seriously and every serious idea treated idiotically. What we make of this is, of course, up to us, but it's hard to avoid at least some level of despair as the traditional authorities lose their grip and before new tools and types of authority have fully settled in. The Internet may not be making me and you stupid, but it sure looks like it's making a whole bunch of other people stupid.

Fourth, we can see—or at least are led to suspect—that every idea is contradicted somewhere on the Web. We are never *all* going to agree, even when agreement is widespread, except perhaps on some of the least interesting facts. Just as information overload has become a fact of our environment, so is the fact of perpetual disagreement. We may also conclude that even the ideas we ourselves hold most firmly are subject to debate, although there's evidence (which we will consider later) that the Net may be driving us to hold to our positions more tightly.

Fifth, there is an odd consequence of the Net's filtering to the front. The old library Acquisitions Committee did its work behind closed doors. The results were visible to the public only in terms of the books on the shelves, except when an occasional controversy forced the filters themselves into the public eye: Why aren't there more books in Spanish, or why are so many of the biographies about men? On the Net, the new filters are themselves part of the content. At their most basic, the new filters are links. Links are not just visible on the Net, they are crucial pieces of information. Google ranks results based largely on who's linking to what. What a blogger links to helps define her. Filters are content.

Sixth, filters are particularly crucial content. The information that the filters add—"These are the important pages if you're studying hypercomputation and cognitive science"—is itself publicly available and may get linked up with other pages and other filters. The result of the new filtering to the front is an increasingly smart network, with more and more hooks and ties by which we can find our way through it and make sense of what we find.

So, filters have been turned inside out. Instead of reducing information and hiding what does not make it through, filters now increase information and reveal the whole deep sea. Even our techniques for managing knowledge overload show us just how much there is to know that escapes our best attempts. There is no hiding from knowledge overload any more.

The New Institution of Knowledge

We are inescapably facing the fact that the world is too big to know. And as a species we are adapting. Our traditional knowledge-based institutions are taking their first hesitant steps on land, and knowledge is beginning to show its new shape:

Wide. When British media needed to pore through tens of thousands of pages of Parliamentarians' expense reports, they "crowd-sourced" it, engaging thousands of people rather than relying on a handful of experts. It turns out that, with a big enough population engaged, sufficient width can be its own type of depth. (Note that this was not particularly good news for the Parliamentarians.)

Boundary-free. Evaluating patent applications can't be crowd-sourced because it would require expertise that crowds don't have. So, when the

US Patent Office was frustrated by how long its beleaguered staff was taking to research patent applications, it started a pilot project that enlists "citizen-experts" to find prior instances of the claimed inventions, across disciplinary and professional lines. That pilot is now becoming a standard part of the patent process.[32]

Populist. IBM has pioneered the use of "jams" to engage the entire corporation, at every level and pay grade, in discussing core business challenges over the course of a few days. From this have come new lines of business—created out of a stew in which the beef, peas, and carrots all have the same rank.

"Other"-credentialed. At the tech-geek site Slashdot.com (its motto is "News for Nerds"), you'll find rapid bursts of argumentation on the geeky news of the day. To cite your credentials generally would count against you, and if you don't know what you're talking about, a credential would do you no good. At Slashdot, a slash-and-burn sense of humor counts for more than a degree from Carnegie Mellon.

Unsettled. We used to rely on experts to have decisive answers. It is thus surprising that in some branches of biology, rather than arguing to a conclusion about how to classify organisms, a new strategy has emerged to enable scientists to make progress together even while in fundamental disagreement.

Together, these attributes constitute a thorough change in the shape of our knowledge-based institutions. To see such changes at work, let's look at two examples, one in business and one in government.

When Jack Hidary built his company, Primary Insight, he looked at the leader in his industry—The Gartner Group—and asked himself what wouldn't it do? Gartner, a consultancy specializing in the business use of technology, hires experts out of information technology industries, gives them a staff, and charges companies in those industries to hear what the specialists have to say. They often become so expert and authoritative that they not only report on industries, they shape them. By operating within the traditional framework of knowledge and expertise, Gartner has built itself into a 1.3-billion-dollar company.[33]

Hidary left his career as a scientist at the National Institute of Health in part because putting scientific papers through the traditional peer-review process had begun to seem frustratingly outdated. There had to be more efficient ways to gather and vet information. So, when Hidary started a company to advise financial fund managers, rather than build a Gartner-esque stable of full-time analysts, with a single lead in each practice area, he made a network of thousands of part-time experts available to every client. This arrangement not only provides a wider range of advice but also means that each financial fund manager is talking with a unique, customizable set of people—a network within the network. That's crucial, explains Hidary, because fund managers' competitive edge is knowing what the other managers do not.

Further, unlike at Gartner, Hidary's expert networks consist of part-time experts who maintain their jobs in their given field. Hidary considers this to be a strength, because removing them from their field "removes their real-world edge." Hidary won't talk about what his clients pay for this service, but it is in six figures. And, he claims his subscription renewal rates are "off the charts."[34] His success comes precisely by contravening much of what we have taken for granted about knowledge. For example:

> Rather than hiring a handful of full-time experts who could be marketed as representing the unique pinnacle of industry knowledge, Hidary has built a network that has strength because of the variety of people in it.

> Rather than amassing content as if it could all be deposited in one human library, he has built a network of people and resources that can be enlarged and deployed at will.

> Rather than treating an expert as a full-time job or a profession unto itself, Hidary insists that his experts have their sleeves rolled up, working in their fields.

Rather than publishing a newsletter, the same for each recipient, Hidary's network is always deployed in uniquely personal ways.

Rather than looking for credentials from authenticating institutions, Hidary's experts earn their credentials from their peers.

If Hidary's Primary Insight represents a new type of institutionalization of expertise, Beth Noveck's experiences reflect a very different sort of institution: the White House.

Noveck, who for the first two years of the Obama administration was the leader of its open government efforts, has a fascinating history as an innovator building communities of experts. In June 2009, Noveck convened fifteen people in the DC boardroom of the American Association for the Advancement of Science, so that the AAAS could combine its expertise with that of the public to provide better advice on issues facing executive-branch agencies. Her goal was to shape a new kind of institution that would enable professionals and amateurs to engage on questions—sometimes driving to consensus but at other times providing a range of options and opinions.

Expert Labs is still young as I write this, but it is already quite remarkable. Two of the most staid and prestigious institutions in America—the White House and the American Association for the Advancement of Science—recognize that traditional ways of channeling and deploying expertise are insufficient to meet today's challenges. Both agree that the old systems of credentialing authorities are too slow and leave too much talent outside the conversation. Both see that there are times when the rapid development of ideas is preferable to careful and certain development. Both acknowledge that there is value in disagreement and in explorations that may not result in consensus. Both agree that there can be value in building a loose network that iterates on the problem, and from which ideas emerge. In short, Expert Labs is a conscious response to the fact that knowledge has rapidly gotten too big for its old containers. . . . [35]

Especially containers that are shaped like pyramids. The idea that you could gather data and information and then extract value from them by reducing them with every step upward now seems overly controlled and wasteful. Primary Insight and Expert Labs respond to knowledge overload, a product of the visibility of the network of people and ideas, by creating knowledge in a new shape: not a pyramid but a *network*.

And not just any network. Knowledge is taking on the shape of the Net—that is, the Internet. Of all the different communication networks we've built for ourselves, with all their many shapes—the history of communication networks includes rings, hubs-and-spokes, stars, and more—the Net is the messiest. That gives it a crucial feature: It works at every scale. It worked back when an online index of the Net fit on a hard drive with half the capacity of a typical laptop today, and it works now that there are a trillion Web pages. There's no practical limit to how much content the Net can hold, and no practical limit to how many links we all can make to filter forward the relationships among that content. For the first time, we can deal with the overload of knowledge without squinting our eyes and wishing we were back in the days when sixteen products with sixteen categories of information counted as too scary for "housewives." At last we have a medium big enough for knowledge.

Of course, the Net can scale that large only because it doesn't have edges within which knowledge has to squeeze. No edges mean no shape. And no shape means that networked knowledge lacks what we have long taken to be essential to the structure of knowledge: a foundation.

2

Bottomless Knowledge

BENNETT CERF BECAME an early television celebrity not because he was the publisher and co-founder of Random House but because he was dapper in his bow tie and had a seemingly infinite supply of anecdotes. Here's one from a collection he published in 1943:

> Cass Canfield of *Harper's* was approached one day in his editorial sanctum by a sweet-faced but determined matron who wanted very much to discuss a first novel on which she was working. "How long should a novel be?" she demanded.
>
> "That's an impossible question to answer," explained Canfield. "Some novels, like *Ethan Frome,* are only about 40,000 words long. Others, *Gone with the Wind,* for instance, may run to 300,000."
>
> "But what is the average length of the ordinary novel?" the lady persisted.
>
> "Oh, I'd say about 80,000 words," said Canfield.
>
> The lady jumped to her feet with a cry of triumph. "Thank God!" she cried. "My book is finished!"[1]

Of course, she got it wrong. But her strategy was impeccable. She asked a genuine expert, got a correct answer, and reached a decision. And, most important, she could stop asking.

The system worked.

And it works not just for foolish women who in the 1940s served as the butt of our jokes. Faced with the fact that there is too much to know, our strategy has been to build a system of stopping points for knowledge. It's an efficient response, well-suited to the paper medium by which we preserved and communicated knowledge.

Let's walk it backward.

It's 1983. You want to know the population of Pittsburgh, so instead of waiting six years for the Web to be invented, you head to the library. The card catalog leads you to an almanac, and the almanac's index points you to the needle of fact in a thousand-page haystack. "Aha, Pittsburgh's population is 2,219,000," you say to yourself, writing it down so you'll remember. The almanac publisher got the information from the US Census. The Census Department sent out hundreds of thousands of folks to knock on doors. They of course had to be trained, and before that a system had to be created for collecting and processing the information they sought. The most recent Census cost $6.9 billion just for 2010, not including the other nine years of operations,[2] or about $20 per American. The almanac cost the library $12.95. The facts in the almanac cost the library less than a penny each.

The economics of knowledge make sense only if, after looking up the population of Pittsburgh in the almanac, people stop looking. If everyone were to say "Well, that may be a pretty good guess, but I can't trust it," and then hire their own census takers to recount the citizens of Pittsburgh, the cost of knowledge would be astronomical. Distrust is an expensive vice.

We have been right to trust almanacs even without investigating how they ensure the reliability of their information. We've presumed that the almanac's editors have collected its information carefully and have processes in place to make sure that it's accurate. If someone were to challenge your assertion about population data, saying "I got it out of the latest almanac" would probably end the contest. The almanac's presumed authority has stopped the argument. The system has worked again.

Of course, if someone you trusted said that you shouldn't have used *Bob's Guesswork Almanac* because it's shoddily put together and full of typos, you might check a different one. And if human lives or large amounts of money depended on the absolute precision of the answer—that is, if the consequences of being wrong were high enough—you might track down the original data from the US Census, or even hire your own census takers. But short of this, you accept the almanac's answer because the very fact that it was professionally published and stocked in your library serves as a credential attesting to its reliability. Such credentials put the "stop" into stopping points. Just as we as a species can't afford to investigate every fact down to its origins, we can't afford to investigate every credential. So, knowledge has been a system of stopping points justified by a series of stopping points. And for the most part, that works very well, especially since the system is constructed so that generally you can proceed to get more information if necessary: You can follow the footnotes, or check the population figures against a second source, without the cost and expense of commissioning your own phalanx of census-takers.

Our system of knowledge is a clever adaptation to the fact that our environment is too big to be known by any one person. A species that gets answers and can then stop asking is able to free itself for new inquiries. It will build pyramids and eventually large hadron colliders and Oreos. This strategy is perfectly adapted to paper-based knowledge. Books are designed to contain all the information required to stop inquiries within the book's topic. But now that our medium can handle far more ideas and information, and now that it is a connective medium (ideas to ideas, people to ideas, people to people), our strategy is changing. And that is changing the very shape of knowledge.

A History of Facts

In 1954, there were so many polio cases in Boston that Children's Hospital had to perform sidewalk triage before an audience of distressed parents seated in idling cars. So, when Jonas Salk's polio vaccine was successfully tested in 1955, he became a hero to an entire generation. But producing the Salk vaccine depended on an earlier breakthrough by John Enders, who with his colleagues discovered in 1948 how to grow the polio virus outside the human body. At the time, viruses were invisible even to the most powerful microscopes. The only way Enders's team could tell if they'd succeeded in growing the polio virus was to inject it into the brain of a monkey and see if it came down with the disease's awful symptoms. Enders's technique made it possible for Salk to develop the vaccine that should have earned him a Nobel Prize. Enders's work did get Enders one of the prizes, in 1954, even before Salk's vaccine had been proven effective.[3]

The chain of knowledge that produced Salk's vaccine followed the most up-to-date medical processes of the day. Yet, in one important way, the path their breakthroughs took was no different from our most ancient way of proceeding. For example, just as surely as we know about viruses and their enemies, the ancients knew that people came in four flavors or "humors": sanguine, choleric, melancholic, and phlegmatic. Each humor was part of a complex conceptual system of organs, bodily fluids, seasons of the year, astrological signs, and treatments. If your humors got out of whack, you might find yourself sent to the local barber for a helpful bloodletting or a purge. (Don't ask.) This knowledge of how the body worked, and how it integrated into its environment, was believed by the Egyptians, Greeks, and Romans, by Muslims, Christians, Jews, and "pagans." For close to 2,000 years, we humans knew the humors were real and extremely important.

Of course, we were wrong. Your bile doesn't depend on your astrological sign and your liver doesn't depend on your personality. But even though we now thoroughly reject the idea of the humors, Galen of Pergamum, Enders of Connecticut, and Salk of New York, born across a stretch of 1,800 years, all accepted and assumed that knowledge works essentially the same way: Knowledge is a structure built on a firm foundation that lets us securely add new pieces—polio myelitis makes sense if you already know about viruses and immune systems. Because we can lay firm foundations, our species gets smarter and better able to survive our virus-laden environments.

Of course, there are vital differences between the houses of knowledge inhabited by Galen and by Enders and Salk. The humor-ists assumed that their foundation was strong and true because it enabled them to draw analogies among all the different realms, from the biological to the social to the psychological to the astronomical. That made sense when we believed that God ordered His universe in the maximally beautiful way, that He gave us minds so we could appreciate His handiwork, and that our minds worked (in His image) by associating ideas. So, to see analogies was to see God's order. Of course, we modern folks don't think analogies are a scientific way of proceeding, or else we'd still think that because some tumors cause veins to swell and look like crabs, cancer must have something to do with the Crab constellation. We believe the firm foundation of knowledge consists not of analogies but of facts. The ancients and we moderns disagree about how to lay a firm foundation, but we firmly believe in foundations themselves.

Facts are facts. It's a fact that polio vaccine is effective and it's a fact that Cancer the constellation has nothing to do with tumors. But the idea that the house of knowledge is built on a foundation of facts is not itself a fact. It's an idea with a history that is now taking a sharp turn.

—

In 2006, former President Bill Clinton wrote an op-ed in the *New York Times* about the legacy of the welfare reform legislation he had championed a decade earlier.[4] His verdict: "The last 10 years have shown that we did in fact end welfare as we knew it, creating a new beginning for millions of Americans." He then supported that assertion:

> In the past decade, welfare rolls have dropped substantially, from 12.2 million in 1996 to 4.5 million today. At the same time, caseloads declined by 54 percent. Sixty percent of mothers who left welfare found work, far surpassing predictions of experts. Through the Welfare to Work Partnership, which my administration started to speed the transition to employment, more than 20,000 businesses hired 1.1 million former welfare recipients. Welfare reform has proved a great success.

At the heart of Clinton's argument is a series of facts. If we wanted to argue against him, we might suggest that those facts are cherry-picked, out of context, or even blatant lies. But to do this, we'd have to offer our own set of facts. We might point out that the poverty level was declining even before Clinton's act went into effect,[5] and that the number of Americans with zero income who receive food stamps has soared to about 6 million because Clinton's bill cut off other sources of cash support.[6] President Clinton would certainly respond with more facts, for we fight facts with facts.

It wasn't always so. In 1816, the British House of Commons debated creating a committee to look into requiring children to be at least nine years old before entering the workforce and limiting their workday to 12.5 hours. This was fiercely opposed by factory owners who were working children as young as six for up to sixteen hours a day. The arguments for and against the law were based on principles and generalities, not facts: "Such a proceeding was libel on the humanity of parents," argued a Mr. Curwen, who believed that parents are the best judges of what's best for their children.[7] Nevertheless, a committee was formed to investigate the situation. Even the experts called by the committee—quite a new practice in itself—were fact-free. Ashley Cooper, Esq., a surgeon, testified that he didn't think children aged seven to ten could work more than ten hours a day without harming their health. "Upon a subject of this kind, one must answer upon general principles," he said, pointing to the need for "air, exercise, and nourishment." Sir Gilbert Blane, M.D., backed up that opinion, although he noted: "I have no experience as to manufactories, and, therefore, my answer must depend on general analogy."[8] General principles may be true, and general analogies may work, but we moderns know that they need to be supported by reams of facts before they can be trusted. Bill Clinton would not have given us just some untested adages.

There certainly were facts before the start of the nineteenth century; it was a fact that the ocean was salty even before humans first tasted it, and it was a fact that polio is caused by a virus even before we had discovered viruses. But only relatively recently have facts emerged as the general foundation of knowledge and the final resort of disagreements.

Indeed, we didn't have a word for facts until a few hundred years ago. When in 400 A.D. Jerome translated John 1:14 ("and the Word was made flesh") into the Latin "et Verbum caro factum est," "factum" meant "that which was done," from "facere" ("to do").[9] The word "fact" entered English in the early 1500s with that meaning, but by the 1600s facts were a narrower class of deeds, as in "He is . . . hanged . . . neere the place where the fact was committed" (1577).[10] Facts were evil deeds, so a murder would have been a fact, but not that the Pyramids are in Egypt. How did we manage so long without a word for what we currently mean by "fact"?

For us moderns, the hardest, most solid of facts are about particulars (there is a rock by the side of the road, there are six chairs around the table), but our ancestors tended to disdain the particular because it comes to us through bodily perception, a capability we share with all animals. For them, knowledge had to be something more than what we learn through our mere senses, because it is such a distinctly human capability of our God-given and God-like soul. Whereas perception sees individual items (this berry, that cat), knowledge discerns what this cat has in common with all other cats that makes it into a cat; knowledge sees its essence as a cat. For our ancestors, knowledge was of universals—not of facts about this or that cat. The idea that knowledge was a slew of facts about particulars would have struck them as a misuse of our God-given instrument.

So what happened in the nineteenth century to make facts the bedrock of knowledge? The path is twisty. Mary Poovey cites the invention in Italy of double-entry bookkeeping, which in the sixteenth century provided a process by which ledger entries could be proved accurate to anyone who, regardless of status, followed the proper procedure.[11] But most historians look to the seventeenth century, when the philosopher and statesman Francis Bacon, seeking to put knowledge on a more certain basis, invented the scientific method. Like Aristotle, he sought knowledge of universals.[12] But he proposed getting to them through careful experiments on particulars. For example, when Bacon wanted to find out how much a liquid expands when it becomes a gas, he filled a one-ounce vial with alcohol, capped it with a bladder, heated the alcohol until the bladder filled, and then measured how much liquid was left.[13] From this experiment on particulars, he was able to propose a theory that applied universally to heated liquids.

Having the particular ground the universal was a remarkable inversion of the traditional approach to knowledge: No longer derived by logical deduction from grand principles, theories would hereafter be constructed out of facts the way houses are built out of bricks.[14]

One more turn and we get to modern facts. When trying to understand a word as basic as "fact," the twentieth-century British philosopher John Austin recommended considering what it is being used in contrast to. Bacon contrasted facts with theories. But the modern sense of facts emerged only when we started contrasting them with people's self-interest: the facts of what cleaning chimneys did to little boys versus the upper class's interest in getting their chimneys clean, which led the rich telling themselves that hard work builds little boys' character.[15] This change in the meaning of facts vaulted facts to the center of the social stage, for that's where we often jostle about coordinating our interests. Facts went from what grounds scientific theories to what also grounds social policy.

As if in a stop-motion video that shows a flower blooming, we can see this rise in the social role of facts occurring within the life span of one great thinker. Thomas Robert Malthus, born in 1766, is best remembered for warning in 1798 that while populations grow geometrically (2, 4, 8, 16 . . .), the food supply required to support them grows only arithmetically (1, 2, 3, 4, 5 . . .). Here's his famous "proof":[16]

First, That food is necessary to the existence of man.

Secondly, That the passion between the sexes is necessary and will remain nearly in its present state. . . .

Assuming then, my postulata as granted, I say, that the power of population is indefinitely greater than the power in the earth to produce subsistence for man.

Malthus doesn't painstakingly assemble facts about population growth and crop yields. Instead, he gives us a logical deduction from premises that he presents as self-evident. From this he goes further: Starvation is inevitable, and therefore those born poor should not be sustained by the government since there isn't enough food to go around anyway. We must raise our moral standards and procreate less.

Malthus spends the rest of the book unfurling a series of bold, unsupported generalizations explaining why the various populations of the earth haven't already grown themselves to the point of starvation. Even when dealing with his own countrymen, he proceeds in a remarkably fact-free fashion. The "higher classes" don't feel the need to marry because of the "facility with which they can indulge themselves in an illicit intercourse."[17] Tradesmen and farmers cannot afford to marry until they're older. Laborers can't afford to divide their "pittance among four or five" and thus do not overpopulate. Servants would lose their "comfortable" situations if they married. The modern researcher would be appalled at these generalizations. What are the average family sizes for each of these classes? What is the average salary of laborers? What are the costs of raising a family of four or five? How common is "illicit intercourse" among the rich, and how does it compare to the rate among the other classes? If Malthus had submitted his book to a sophomore college class for his final paper, he would be sent back to a course on Remedial Methodology to rewrite it.

Throughout his life, Malthus did indeed rewrite his masterwork. By the time he published the sixth and final edition in 1826, it was dense with facts, statistics, and discussions of the validity of contrasting studies. He compared mortality rates in different regions, explained anomalous statistical results, and in general behaved like a fact-based modern researcher. The change clearly was due in part to the availability of more facts. But facts became more available because their status had grown. Fact-based knowledge was arriving, spurred on in part by Malthus's own work, and driven by reformers who wanted society to face the miserable reality of England's working class and poor. In that political struggle, facts got contrasted with interests, and not just with theory.

And interests could not have been much further apart than they were at the beginning of the nineteenth century, when the upper class, secure that its position was part of the divine plan, felt no compunction about setting children to work in factories or sending boys as young as five up chimneys as narrow as seven inches square. In 1819, when the British House of Commons considered a new bill that would have kept children younger than fourteen out of chimneys, a member of the House named Mr. Denman argued that it was better that the boys be gainfully employed than that they engage in "the fraud and pilfering which was now so common among boys of tender age."[18] A Mr. Ommaney agreed, for the chimney sweeps he had seen were "gay, cheerful, and contented." And, added Mr. Denman, there was no other reliable way to clean the really small flues for which little boys are perfectly shaped.

Yet, the tide was turning. The proponents of the reform bill countered Mr. Ommaney's vision of "gay, cheerful, and contented" chimney sweeps with factual evidence from physicians that these boys "exhibited every symptom of premature old age." The proponents won. The fact of the misery of these children overcame the old assumption that the poor are poor because they deserve to be—a belief that conveniently supported the self-interest of Mr. Denman, Mr. Ommaney, and the rest of the old guard.

The chimney sweep bill was just one sign, not a turning point. The triumph of facts was gradual. But it was greatly enabled by a single thinker who provided the intellectual framework for basing policies on facts rather than on moral assumptions or the interests of those in power. Jeremy Bentham was "one of the few great reformers to be appreciated in his own lifetime."[19] By the time he was four, he was learning Latin and Greek. By the time he was five, he was known as "the philosopher." At age thirteen he was admitted to Oxford and trained to become a lawyer like his father. But he was too curious to stay within any one domain and became, among other things, a philosopher best remembered for the principle of utility: Since pleasure and pain are equal motivators for all people, Bentham declared, the ultimate criterion for evaluating an action was whether it would result in "the greatest happiness of the greatest number." Bentham thus gave us a new way of evaluating social policies, a type of bookkeeping committed to improving the ledger overall.

Applied to government, Bentham's ideas were radical. Suppose the happiness of Mr. Denman and Mr. Ommaney did not count for more than the happiness of the boys sweeping their chimneys. Suppose government policies should be guided by a pragmatic sense of what works to increase the overall happiness. If so, the government would first need to survey what life was actually like for all of its citizens. It would need to base policy on *facts*.

But that in turn required the use of a tool only recently gaining credence: statistics. The word itself only entered English around 1770, coming from a German word for information about the state (which explains the "stat" part of the word).[20] Statistics were meant to be independent of opinions and conclusions. They became the way the social reform movement argued against the conclusions that interested parties might have preferred.[21]

In the 1830s, statistics gave Bentham's ideas a method, and Parliament, partially due to Bentham's influence, soon commissioned reports on poverty, crime, education, and other social concerns. Distributed in "blue books" rich with anecdotes, interviews, and statistical tables, these reports put parliamentary debate of social issues on a factual basis, even if their statistical methods were not up to modern standards. Blue books also provided material for popular novels that further advanced the social reform movement.[22]

One of the most famous of these novelists of social reform was, of course, Charles Dickens. But in London, blue books were flying into Parliament and off the shelves at such a pace that by 1854 Dickens was part of a backlash against the whole fact-based approach. In *Hard Times,* Dickens has the schoolmaster Thomas Gradgrind exhort his students: "'Fact, fact, fact!'"[23] The unsympathetic Mr. Gradgrind tells his students that they must not decorate their future homes with carpets with floral designs. "'You don't walk upon flowers in fact; you cannot be allowed to walk upon flowers in carpets. You don't find that foreign birds and butterflies come and perch upon your crockery; you cannot be permitted to paint foreign birds and butterflies upon your crockery."[24] Facts, for Dickens, stood in contrast to imagination and art, and were too dry a way to understand human life.

Dickens made it clear to his readers that the rise of facts to which he objected was coming out of the political sphere. Gradgrind tells his students: "'We hope to have, before long, a board of fact, composed of commissioners of facts, who will force the people to be a people of fact, and of nothing but fact.'"[25] In case the political reference wasn't clear enough, Dickens tells us that Mr. Gradgrind's room had an "abundance of blue books." Says Dickens: "Whatever they could prove (which is usually anything you like), they proved there. . . . In that charmed apartment, the most complicated social questions were cast up, got into exact totals, and finally settled. . . . As if an astronomical observatory should be made without any windows, and the astronomer within should arrange the starry universe solely by pen, ink, and paper."[26] Dickens had enormous sympathy for the poor, having worked in a shoe polish factory at the age of twelve and having watched his father taken away to debtors' prison.[27] But facts in blue books didn't reveal the truth. For that you needed to understand in depth and compassionately the lived plights of social unfortunates, the way we do when reading a novel. What a coincidence!

Even with the incredibly popular Mr. Dickens railing against overreliance on facts, they continued their rise to prominence. With the invention of "fact-finding missions," facts became the basis for resolving international disputes. Indeed, it's hard to believe that they're a modern invention, but the first mention of a fact-finding mission in the *New York Times* occurred in 1893 when President Grover Cleveland sent someone to investigate the dethroning of Hawaii's last queen, Lili'uokalani.[28] These commissions became a normal procedure only after the Hague conference created its first fact-finding mission in 1904, when five countries jointly investigated the mistaken sinking of an English trawler by Russia's Baltic Fleet.[29] Russia paid England reparations for the so-called Dogger Bank Incident, and for the first time, an international dispute was settled by uninvolved—disinterested—countries working to establish the facts and nothing but the facts. By the

1920s, fact-finding missions had become a normal and accepted part of how countries tried to settle their problems[30]—possibly because the 16.5 million deaths in World War I showed that the other popular method of settling disputes didn't work so well. These days, if something large enough goes wrong, we create a fact-finding mission as if this were a natural and age-old way of proceeding.

Over the course of two hundred years, facts came a long way. From the opposite of theories, to the opposite of self-interest, to the way unfriendly countries avoid war, facts became the elemental truths about the world—truths that are true regardless of what we may think or want to believe. Journalists gathered facts, almanacs aggregated facts, board games quizzed us on them, experts predicted entire baseball seasons based on previous seasons' facts, governments prepared to deploy Armageddon's weapons based on cold-hearted assessments of facts. Facts had hit rock bottom, which is exactly where we wanted them.

Darwin's Facts

One day in the 1850s Henry David Thoreau observed a bird he hadn't seen before "flapping low with heavy wing." As the bird flew overhead, Thoreau caught sight of two spots on the bottom of its wings and realized it was a kind of gull. "How sweet is the perception of a new natural fact!"[31] Thoreau chirruped. A new fact had been uncovered: This particular bird was a gull. Thoreau's fact is in the fact's most basic form: Some *this* is a *that*.

Yet Thoreau's identification of that bird wasn't the sort of fact that does the heavy-lifting of knowledge. It did not advance our knowledge of gulls, of wings, or even of spots in any appreciable way. Thoreau was not that ambitious. As Ralph Waldo Emerson lamented in his eulogy of his friend, "instead of engineering for all America, he was the captain of a huckleberry party."[32]

While Thoreau was picking huckleberries, Charles Darwin was spending seven years intently exploring the small world of Cirripedia—barnacles. The two resulting dry and difficult volumes—so little like his masterful *On the Origin of Species* published just a few years later in 1859—are careful recitations of facts that together describe the little creatures in unrelenting detail. But they lead up to a *this* is a *that* far more consequential than Thoreau's. It is a classic example of how fact-based knowledge has worked, and has worked so well that it can be worth spending seven years going to the family dinner table smelling of dead crustaceans.

Darwin's work on barnacles began with the accidental discovery of a small, persistent fact. In 1835, before he had formulated his great theory, Darwin was a young man sailing on the *Beagle,* exploring the small variations in the plants and animals of the Galapagos Islands. There he discovered tiny barnacle parasites inside the shell of a mollusk—highly unusual for creatures that usually attach to rocks. Examining the parasites more closely, he found tiny larvae that looked surprisingly like crustacean larvae. Mollusks and crustaceans were classified as separate species, so why would one produce larvae of another? Darwin filed the question away until 1846, when for the next seven years it fully absorbed him.

To call this work painstaking would be to underestimate it vastly. But the minutiae of his work with needle-like dissection tools and magnifying equipment was guided by a grand theory. The idea that organisms evolve by small steps led Darwin to look for continuities among them. Accordingly, he probed hermaphroditic barnacles and found male organs that were so "unusually small" that he would not "have made this out, had not my species theory convinced me, that an hermaphrodite species must pass into a bisexual species by insensibly small stages."[33]

Darwin's first volume on barnacles is 370 pages long but is really about a single fact: Barnacles are crustaceans. This fact could not have been uncovered by thinking about the world while sitting on the banks of Walden Pond, and bringing your laundry home for your mother and sister to wash (as Thoreau did).[34] Darwin's fact required a trip from England to the Galapagos, the close inspection of a mollusk with parasites, the acquisition of multiple collections of specimens, seven years of exacting dissections, and a world-changing theory of animal origins.

How sweet indeed is the perception of a new natural fact.

Now flash forward to the present and ask Hunch.com for help in any matter of taste. What city should I visit? What character should I dress up as for Halloween? What Chinese vegetable should I cook tonight? Hunch.com will supply statistically significant answers based upon the rippling, overlapping similarities among all its users. For this to work, Hunch has to know lots about its users—so much that asking them to fill in a typical profile ("Favorite type of music," "Politics: left, middle, or right") would not even begin to suffice. Hunch is looking for a sort of fact that would have confounded Darwin, Thoreau, and most of us just a few years ago.

When I first visited the site, wondering what movie I should see that evening, Hunch asked me a series of questions that had nothing to do with movies. Do I store my drinking glasses right side up? Would I prefer to wear running shoes, boots, or sandals? When I throw out a sheet of paper, do I crumple it? Have I ever touched a dolphin? I have answered a total of 334 such questions since I began using the site, primarily because answering them is surprisingly fun. On the basis of my answers, Hunch recommended the movies *28 Days, Casablanca, The Fugitive,* and *The Big Lebowski.* Hunch.com has got my number.

It got that number by analyzing my answers in the context of millions of answers given by other users. The analysis is purely statistical, in a way that the nineteenth-century scientists and statisticians would not have foreseen. The analysis is not in support of a theory and it produces no theory. Hunch has no idea why people who, say, prefer to wear sandals on a beach and who have not blown a dandelion in the past year might like those four movies. It doesn't have a hypothesis and it doesn't have a guess. It just has statistical correlations.

Hunch's facts—how I store my drinking glasses and whether I've recently blown on a dandelion—are the opposite of Darwin's facts:

Darwin's facts were hard-won. He spent seven years establishing that barnacles are crustaceans. At Hunch, you can answer 12 questions a minute. The average user has answered about 150 of them. Facts are fast and fun at Hunch.

Darwin's facts were focused on a particular problem: understanding what sort of critter the barnacle is. Hunch's facts are purposefully unconstrained. One moment you're answering a question about your favorite ABBA song and the next you're declaring whether you consider Russia to be part of Europe. Hunch needs answers to be spread wide and thin in order to generate useful results.

Darwin's facts together cover some finite topic. In 370 pages, Darwin goes through all of the relevant facts about the three types of barnacles, and nails his argument. Granted, that's a lot of pages and even more facts, but it has a beginning and an end. It fits between covers. Hunch's facts don't "cover" anything. In its first seven months, the site gathered over 7,000 different questions, almost all of them from its users. The only stopping point is when you're tired of answering silly questions. And even then you can always go back for more.

Darwin's facts existed before he discovered them. Hermaphroditic barnacles had tiny male organs before Darwin peeked at them. It's not nearly as clear whether Hunch is uncovering or generating facts. It's a fact about me that I would prefer cotton candy to a shoe shine, but since I had never before considered the comparison, it feels like a fact that didn't quite exist before someone asked. If my never having touched a dolphin counts as a fact, then so must the fact that I have never touched a Klingon, or a purple lime, or a blue lime, or a plaid lime—an infinite series of facts that didn't exist until someone asked.

Darwin's facts emerged because he had a theory that guided him. Otherwise, why care about hermaphroditism in barnacles? Hunch doesn't know why your preference in salty snack foods might help predict your favorite type of poker, and it doesn't care.

Finally, when Darwin noticed a parasitic barnacle in the Galapagos, it was a fact worth remembering only because he assumed—correctly—that this individual barnacle was representative of a species. When he writes in Volume 1 that "in *L. anatifera* alone, the uppermost part of the peduncle is dark,"[35] he's referring not to an individual barnacle's peduncle but to the species' pedunculosity. For Darwin, the facts worth noting are the ones that apply to more than one individual. Exactly the opposite is the case at Hunch. "Are you an air-breather?" is not a helpful question at Hunch because all of its mammalian users will reply the same way.

Now, Hunch is not producing results on the order of Darwin's barnacle studies or his *Origin of Species*. Nor do Hunch's facts replace the need for Darwinian-style facts (although in Chapter 7 we will see how science is using some of Hunch's basic techniques). Hunch is doing something useful—helping you find the next movie to see or the right wedding gift to buy—but it's not making any serious claim to producing eternal knowledge. It's just about, well, hunches.

Nevertheless, Hunch is a trivial example of a serious shift in our image of what knowledge looks like. Darwin's facts were relatively scarce both because they were hard to obtain—seven years dissecting barnacles—and because they were hard to get published. Some facts are still so hard to obtain that multi-country consortia have to spend billions of dollars building high-energy particle colliders to get them to show their quantum-scale faces. But our information technologies are precisely the same as our communication technologies, so learning a fact can be precisely the same as publishing a fact to the world. The Internet's abundant capacity has removed the old artificial constraints on publishing—including getting our content checked and verified. The new strategy of publishing everything we find out thus results in an immense cloud of data, free of theory, published before verified, and available to anyone with an Internet connection.

And this is changing the role that facts have played as the foundation of knowledge.

The Great Unnailing

The late, revered senator from New York, Daniel Patrick Moynihan, famously said, "Everyone is entitled to his own opinions, but not to his own facts."

Perhaps this is what President Barack Obama had in mind when he took as his first executive action the signing of the "Transparency and Open Government" memorandum, requiring executive-branch agencies to "disclose information rapidly in forms that the public can readily find and use."[36] Two months later, Vivek Kundra, Obama's pick for the new post of federal Chief Information Officer, announced plans to create a site—Data.gov—where executive-branch agencies were required to post all their nonsecret data so the public can access it—everything from requests received by the Department of Agriculture for permits for genetically engineering plants to the National Cemetery Administration's customer satisfaction surveys. When Data.gov launched, it had only 47 datasets. Nine months later, there were 168,000,[37] and there had been 64 million hits on the site.[38]

Obama's executive order intended to establish—to use a software industry term—a new default. A software default is the configuration of options with which software ships; the user has to take special steps to change them, even if those steps are as easy as clicking on a check box. Defaults are crucial because they determine the user's first experience of the software: Get the defaults wrong, and you'll lose a lot of customers who can't be bothered to change their preferences, or who don't know that a particular option is open to them. But defaults are even more important as symbols indicating what the software really is and how it is supposed to work. In the case of Microsoft Word, writing multi-page, text-based documents, and not posters or brochures, is the default. The default for Ritz crackers, as depicted on the front of the box, is that they're meant to be eaten by themselves or with cheese.[39]

Before Obama's order, most government data was by default unavailable to the public. The Environmental Protection Agency used to release the results of its highway mileage tests, but not the data such tests were based on. After the new default went into effect, at the EPA's data site—FuelEconomy.gov—you can download a spreadsheet of mileage testing information that will tell you not just that a 2010 Prius gets 51MPG in the city but also that the average annual fuel cost should be $780 and it's got a "Multipoint/sequential fuel injection" system; there's even information about hydrogen fuel cell cars that don't yet exist.[40] Advocates of open government hope that changing the default will make it easier to hold the government accountable, and will spur the development of new software applications that make use of those data, the way US Geological Survey and Census Bureau data have been put to unexpected uses in the past.

The agency-wide change in default effected by Kundra on behalf of the Obama administration was intended to signal something important about the role and nature of government, but it also tells us something about the changing role and nature of facts. FuelEconomy.gov may give us one hundred categories of data, but there are no columns for the ambient temperature, the pounds per square inch of the air in the tires, or the pull of the moon on the day the road tests were done, all of which might have some small effect on the data. We know that there could be another hundred, thousand, or ten thousand columns of data, and reality would still outrun our spreadsheet. The unimaginably large fields of data at Data.gov—we are back to measuring stacked *War and Peaces*—do not feel like they're getting us appreciably closer to having a complete picture of the world. Their magnitude is itself an argument against any such possibility.

Data.gov and FuelEconomy.gov are not parliamentary blue books. They are not trying to nail down a conclusion. Data.gov and the equivalents it has spurred in governments around the world, the massive databases of economic information released by the World Bank, the entire human genome, the maps of billions of stars, the full text of over 10 million books made accessible by Google Books, the attempts to catalog all Earth species, all of these are part of the great unnailing: the making accessible of vast quantities of facts as a research resource for anyone, without regard to point of view or purpose. These open aggregations are often now referred to as "data commons," and they are becoming the default for data that has no particular reason to be kept secret.

This unnailing is perceived by some people as potentially dangerous. For example, not long after Data.gov was begun, open government proponents were surprised to read an article in a liberal journal by one of their great advocates, Lawrence Lessig, about open data's downside. In the article, titled "Against Transparency," Lessig warned that making available oceans of uninterpreted data may lead politically motivated operatives to draw specious connections: Every time a candidate accepts funds from a lobbying group and votes for a bill that the lobbyists favor, such operatives could claim that this is proof that the candidate is corrupt. Thus, this unnailed data could further enable an accusatory culture. (Lessig's article proposes reducing citizens' cynicism by reforming the United States' campaign finance process.)[41]

This is a second irony of the great unnailing: The massive increase in the amount of information available makes it easier than ever for us to go wrong. We have so many facts at such ready disposal that they lose their ability to nail conclusions down, because there are always other facts supporting other interpretations. Let's say I gather data on global climate change from World Resources Institution's collection of information from two hundred countries, and you grab some different data from Fauna Europaea's database of the distribution of species. You don't like my conclusion? Within a couple of seconds, you can fill your own bag with facts.

Our foundations are harder to nail down than they used to be.

—

Facts have changed not only their role in arguments but their own basic shape. We can distinguish three phases in the recent history of facts (although the division is much messier than that).

First, there was the Age of Classic Facts, represented by Darwin with a dissecting kit and by parliamentary blue books. These facts were relatively sparse, painstakingly discovered, and used to prove theories.

Then, in the 1950s we entered the Age of Databased Facts, represented by punchcards stacked next to a mainframe computer. We thought we had a lot of information then, but it would have taken just under 2 billion cards to store what's on a rather wimpy 200-gigabyte hard drive on a laptop—a stack about 300 miles high.[42] So, of course the databases of the time had to strictly limit the amount of information they recorded: the employee's name, date of birth, starting date, and Social Security number, but not hobbyist skills or countries lived in. The Age of Data still conformed to our ancient strategy for knowing the world by limiting what we know—a handful of fields, chosen and organized by a handful of people.

Now, in the Age of the Net it makes sense to talk about networked facts. If classic facts and databased facts are both taken as fundamentally isolated units of knowledge, networked facts are assumed to be part of a network. Networked facts exist within a web of links that make them useful and understandable. For example, in the days of print, the tables of data in a scientific article were tiny extracts from masses of facts and data that themselves were not published. Now, on the Internet, scientific journals are increasingly hyperlinking from the data in their articles to the databases from which they are drawn. For example, when an article in the journal *Public Library of Science Medicine*[43] examines "the predictors of live birth" in *in vitro* fertilization by analyzing 144,018 attempts, it links to the UK open government site where the source data—"the world's oldest and most comprehensive database of fertility treatment in the UK"—is available.[44] The new default is: If you're going to cite the data, you might as well link to it. Networked facts point to where they came from and, sometimes, where they lead to. Indeed, a new standard called Linked Data is making it easier to make the facts presented in one site useful to other sites in unanticipated ways—enabling an *ad hoc* worldwide data commons. Key to Linked Data is the ability for a computer program not only to get the fact but to ask the resource for a link to more information about the context of the fact.[45]

Facts have become networked because our new information infrastructure happens also to be a hyperlinked publishing system. If you're going to make a fact visible, it's so easy to link it to its source that you'll need some special justification not to do so. But our new network doesn't just unify our information and publishing systems. It also integrates us with other people. In the EPA's database of car mileage, the Prius's 51 miles per gallon is just a number. Once that fact is embedded in a labeled table, it becomes meaningful. Then, when someone posts it on a page, it picks up more meaning. Whatever point that page is making—Prius's mileage is great, isn't great enough, is a sham—it's quite likely that somewhere another page links to that one to argue the other way. Thus, the networked datum "51" points back to a traditional database, but also points ahead into the unruly context of networked discussion. This makes our ordinary encounter with facts very different from what it used to be. We don't see them marching single-file within the confines of an argument contained within a blue book, a scientific article, or a printed tome. We see them picked up, splatted against a wall, contradicted, torn apart, amplified, and mocked. We are witnessing a version of Newton's Second Law: *On the Net, every fact has an equal and opposite reaction.* Those reactive facts may be dead wrong. Indeed, when facts truly contradict, at least one of them has to be wrong. But this continuous, multi-sided, linked contradiction of every fact changes the nature and role of facts for our culture.

When Daniel Patrick Moynihan said, "Everyone is entitled to his own opinions, but not to his own facts," what we heard was: Facts give us a way of settling our disagreements. But networked facts open out into a network of disagreement. We may miss the old Age of Classic Facts, but we should recognize that its view of facts was based not in fact but in the paper medium that published facts. Because of the economics of paper, facts were relatively rare and gem-like because there wasn't room for a whole lot of them. Because of the physics of paper, once a fact was printed, it stayed there on the page, uncontradicted, at least on that page. The limitations of paper made facts look far more manageable than they seem now that we see them linked into our unlimited network.

Of course, there are important domains where facts play their old role, and we would not want it otherwise—lots of lives were saved because Jonas Salk had a methodology for proving that his vaccine worked. The Net lets us find that fact, and explore its roots. And yet, the longer you are on the Net, the more fully you realize that in so many areas, facts fail at their old job. The people who think vaccines cause autism, the ones who still think Barack Obama was born in Kenya, the ones who think the government is hiding proof that aliens walk among us, they all have more facts than ever to prove their case. And so do those who think (as I do) that those beliefs are crazy. We see all too clearly how impotent facts are in the face of firmly held beliefs. We have access to more facts than ever before, so we can see more convincingly than ever before that facts are not doing the job we hired them for.

Let me stress that the old role of facts does not vanish from the Net. Scientists still establish facts as in the old days, thankfully. Policy debates continue to try to ground their conclusions in facts, although as always there are fierce arguments over which facts are relevant and what to make of them. And, importantly, the realm of commoditized facts—facts that a large community of belief accepts as not worth arguing about—is growing, as is access to those facts: Anyone with a Web browser can get a figure for the population of Pittsburgh that for almost all conceivable purposes will count as reliable enough. But push on a fact hard enough, and you'll find someone contradicting it. Try to use facts to ground an argument, and you'll find links to those who disagree with you all the way down to the ground. Our new medium of knowledge is shredding our old optimism that we could all agree on facts and, having done so, could all agree on conclusions. Indeed, we have to wonder whether that old optimism was based on the limitations inherent in paper publishing: We thought we were building an unshaken house based on the foundation of facts simply because the clamorous disagreement had no public voice.

In short, while facts are still facts, they no longer provide the social bedrock that Senator Moynihan insisted on.

And, by the way, there is no solid certainty that Senator Moynihan ever actually said "Everyone is entitled to his own opinions, but not to his own facts." It might have been a variant, such as "You are entitled to your own opinion, but you are not entitled to your own facts." It might actually have been James Schlesinger who said it. That Senator Moynihan ever uttered that phrase simply is not a known fact.

I learned that on the Internet.[46]

3

The Body of Knowledge

An Introduction to the Rest of the Book

THE BIBLE'S ABRAHAM, in a land of idol worshippers, thought knowledge is what you see beyond what is before your eyes. The Athenians thought it was an opinion that is true and that we have good reason to believe. Descartes thought it was that which you could not under any imaginable circumstances be wrong about. Scientists have thought it is that which well-designed, repeatable experiments enable us to have confidence in. There is no shortage of definitions of knowledge. And there is no overall agreement.

But there have been a few characteristics of knowledge that have persisted throughout the ages in the West: First, knowledge is a subset of belief. We believe many things, but only some of them are knowledge.

Second, knowledge consists of beliefs that we have some good reason to believe, whether it's because we've done experiments, because we've proved them logically, or because God revealed them to our people.

Third, knowledge consists of a body of truths that together express the truth of the world.

The first two are affected by the networking of knowledge. The third is being erased. We are losing knowledge's body: a comprehensible, masterable collection of ideas and works that together reflect the truth about the world. In field after field we've witnessed the idea of a "canon" falling. The idea that there is such a thing as "the news" that could possibly fit into a daily newspaper or newscast, that there are agreed-upon Great Works of Literature that make one literate, that there's a reasonable way to pare an encyclopedia down to a mere 65,000 entries, that we even know what constitutes a civilization—all of these notions have been under attack for a couple of generations now. The Internet is sealing the deal.

On the one hand, we'll miss knowledge's body. We like picturing knowledge as a collection of truths that have made it past wise custodians. The collection—iconically, a library—is always growing as we learn more and more. We learn from it and, perhaps, our own work will add to it.

On the other hand, day to day, we won't much miss it. We'll still have the sort of knowledge that lets us operate in the world. Bus schedules will continue to be somewhat accurate. We'll still have the movie reviews we rely on, whether they're in printed newspapers or blogs. We'll still have the facts that we take for granted—two plus two equals four, Albany is the capital of New York—and that sort of knowledge will be easier to find than ever. Marketers will still claim that a shirt is "as soft as a monsoon breeze" and we'll know that the shirt is not quite as soft as that. What's at issue is Knowledge with a capital K, the sort of tested, authorized truths that get carefully placed in the Pantheon of Knowledge, whether they are the principles of science, the broad "irrefutable" generalizations about the nature and aims of human life, or the foundational framing of how the pieces go together. Truths will remain true, but we are losing the sense that we know how to build a Pantheon that is certain, consistent, agreed upon, and much smaller than the universe itself.

So, imagine for a moment that we give up on the idea that we could ever figure out which carefully selected statements are so beyond dispute and so important that they ought to be admitted into the Big Book of Human Knowledge. We would still have knowledge as an important type of belief, one that we have some reason to believe is very likely true. We would still be able to sort through ideas, assigning them various degrees of credibility, from the axiomatic certainty that two plus two does equal four, to the demonstrable truth that boiling liquid water turns some of it into steam, to the actionable likelihood that cheating on taxes will get you in trouble, to the arguable hypothesis that asking companies to be nice will slow global climate change. We'll still have facts. We'll still have experts. We'll still have academic journals. We'll have everything except knowledge as a body. That is, we'll have everything except what we've thought of as knowledge. What would we miss?

This is not a mere thought experiment. It is what the Internet is doing to knowledge. The Internet simply doesn't have what it takes to create a body of knowledge: No editors and curators who get to decide what is in or out. No agreed-upon walls to let us know that knowledge begins here, while outside uncertainty reigns—at least none that everyone accepts. There is little to none of the permanence, stability, and community fealty that a body of knowledge requires and implies. The Internet is what you get when everyone is a curator and everything is linked.

Traditional knowledge is what you get when paper is its medium. There is nothing mystical about this. For example, if your medium doesn't easily allow you to correct mistakes, knowledge will tend to be carefully vetted. If it's expensive to publish, then you will create mechanisms that winnow out contenders. If you're publishing on paper, you will create centralized locations where you amass books. The property of knowledge as a body of vetted works comes directly from the properties of paper. Traditional knowledge has been an accident of paper.

In the remainder of this book we will follow a train of thought that begins with the hypothesis—for which there is increasing evidence—that in a networked world, knowledge lives not in books or in heads but in the network itself. It's not that the network is a super-brain or is going to become conscious. It's not.* Rather, the Internet enables groups to

---

*I'm leaving this as an unsupported idea because it's not the point of this book. I here merely intend to prevent readers from misinterpreting me in a particularly tempting way. Note also that this footnote itself is an example of two of the problems with long-form thought discussed in Chapter 6: anticipating imagined objections and keeping readers on the bus rather than exploring interesting side roads. (That side road would have argued that thought emerges from flesh, and is not a purely formal process. Interested readers should go out on the Net to explore the concept of embodied thought.)

develop ideas further than any individual could. This moves knowledge from individual heads to the networking of the group. We still need to get maximum shared benefit from smart, knowledgeable individuals, but we do so by networking them. *Chapter 4 is on the networking of expertise.*

But the Internet by its nature contains much diversity and many, many disagreements. We need to explore how to know in a world where people don't—and won't—agree about anything. *Chapter 5 is on the importance and limits of diversity.*

Hyperlinks challenge the traditional way of putting ideas together in a page-turning sequence. Yet we still need to put ideas together in ways that lead us to conclusions. *Chapter 6 is on long-form thinking (books) versus hyperlinked webs.*

Then we put these ideas to the test by looking at two areas where it seems that knowledge has to settle down, shake off the constant disagreements and bickering on the Net, and get real. *Chapter 7 looks at science in the Age of the Net.* How is our most reality-based discipline managing in the new linked chaos?

Then we look at what happens to knowledge when it has to guide action. *Chapter 8 is on decision-making and leadership in a networked world.*

Finally, we inevitably want to know if the networking of knowledge is a good thing or a bad thing. The terms of the question may be terribly ill-defined, but we still want it answered. *Chapter 9 is about what we have to do to make the network a better infrastructure for knowledge.*

All of these chapters explore doing the job of knowing now that the limitations of knowledge's old medium are lifting. Even if the smartest person in the room is the room itself, the room does not magically make all who enter it smart. We need to understand what of the old is worth holding on to, and what limitations of the new technology are going to trap and tempt us. A new strategy for knowing our world is emerging, but we are not passive in its arrival.

4

The Expertise of Clouds

A Brief History of Experts

On February 3, 1986, President Reagan issued Executive Order 12546, establishing a Presidential Commission to report on why the Space Shuttle *Challenger* had blown up five days earlier, one minute and thirteen seconds after it took off.[1]

The report begins with a calm, factual recounting of the disaster in intervals of a thousandth of a second: the first puff of gray smoke at 0.678 seconds; the "continuous, well-defined plume" of flame at 59.262 seconds; the "circumferential white vapor pattern" at 73.124 seconds; the *Challenger* "totally enveloped in the explosive burn" milliseconds later. The report concludes with recommendations in nine areas to remedy engineering faults, procedural flaws, and political pressures on NASA.

It is easy to see why this report is held in such high regard. The commission was headed by former Secretary of State William Rogers and included generals, physicists (including Richard Feynman), astronauts (including Sally Ride and Neil Armstrong), test pilots (Chuck Yeager), and rocket scientists. It took a broad look at the causes of the failure and produced an evidence-based document that led to needed improvements in NASA's processes. The report saved lives. And it did so by embodying the very best of traditional expertise: A relative handful of highly trained and credentialed experts came together, followed a careful process, agreed on conclusions, wrote them down, and published them.

When Chaucer in the fourteenth century wrote in *Troilus and Criseyde* that those who are "expert in love" advise that it helps a man to have an opportunity to reveal his woe, to be expert was simply to be experienced.[2] The idea that being an expert could be a full-time, paid job rose with our culture's increasing belief in science as a guide to social policy.

Some historians trace the rise of professional experts to a meeting held six months after the end of the Civil War,[3] when one hundred reformers in various fields met in the Massachusetts State House and created the American Association for the Promotion of Social Science to advise their local communities and states about fixing everything from education to urban poverty, all based on the latest scientific research.[4] By the early 1900s, experts wielding "scientific management" techniques pioneered by Frederick Wilson Taylor—immortalized as the man with a clipboard and a stopwatch, timing the movements of workers—were sweeping through field after field.[5] Even the home was now subject to the work of experts; as Ellen Swallow Richards, the founder of home economics and the first woman to get an engineering degree from MIT, wrote: "The work of homemaking in this scientific age must be worked out on engineering principles and with the cooperation of trained men and trained women."[6]

Experts as full-time professional knowers needed professional institutions to support them. The first of these, the Brookings Institution, was founded in 1916, to provide policy advice to the government. By the 1950s, the Defense Department was relying on the RAND Corporation to help figure out questions of global life and death, including how nuclear war might be waged "successfully" and thus what types of bombs to build. RAND (the name comes from "Research and Development") gave us our modern image of the expert, and he looked like Herman Kahn. The egg-shaped Kahn made a career by (as the title of his best-selling book put it) Thinking the Unthinkable: how to win a nuclear war. This hyper-rational approach could recommend a nuclear arms strategy because it would kill "only" 10 million people. Kahn entered public awareness as an egghead cheerfully detached from the carnage he contemplated, and was ridiculed in *Dr. Strangelove, Or How I*

Learned to Stop Worrying and Love the Bomb, even as he wielded considerable influence on the Kennedy administration. Experts became cemented in our heads as highly rational, unswayed by personal and political considerations, and sometimes out of touch with lived reality.

In 1970, there were about two dozen think tanks. Today, there are over 3,500 worldwide, with about half of them in the United States.[7] Our government has now relied on think tanks and their experts for a hundred years.

As the influence of think tanks on government policy increased, the cult of expertise spread to every corner of our culture. For example, after World War II, parents began relying on "parenting" experts, even though the experts disagreed among themselves. When should you potty-train your child? What are the effects of day care on very young children? What does a mild slap on the tush actually teach a child? A 2006 book that reviewed the advice of five top experts found little agreement even on such basic issues.[8] Still, we continue to assume that it's not only possible to be an expert at raising children but that experts can transfer that skill to us, often with just a few cogent mottoes.

Bringing smart people together is an ancient and effective technique for developing ideas. The Net also lets smart people connect and communicate. But the Net brings people together in new and occasionally weird configurations—a weirdness that is now being reflected in how expertise works. . . .

From Crowds to Networks

If we were to diagram the way we think about the efficiency of the various ways we organize ourselves, until the past few years it would look roughly like a pyramid. At the bottom are mere crowds milling with multiple aims, like the crowds that pass by on a spring day in a city. As we go up the pyramid, the size of the social groups gets smaller, and their degree of organization increases. It's for sure not a perfect pyramid—we can easily think of groups that are large and well-structured, or small and poorly structured—but the narrowing of the pyramid is based on the general truth of the observation that there is a type of social entropy, a waste of energy, when you try to get efficient results out of a large group. The organization can put in controls so that the whole sprawling collection of people works together, but it takes a lot of energy. Just ask your local army or multinational corporation.

This informal diagram of a social pyramid expresses a definite attitude. When we call people a "crowd" we mean that there is no additive social value to aggregating them. Worse, crowds have been associated with soul-deadening conformity and alienation in the age of mass communications. For example, in their 1950 best-seller *The Lonely Crowd*,[9] David Riesman, Nathan Glazer, and Reuel Denney argued that America's corporate culture was creating a generation of sheep who would conform in order to be socially approved. Even worse than a crowd is another social form at the bottom of the pyramid: a mob. A mob is a crowd stirred to the basest of actions. For example, an anti-draft crowd in New York City in 1863 turned into a mob when it threw stones, started fires, and looted. Before too long, it had lynched black men and torched the Colored Orphan Asylum on Fifth Avenue.[10]

So, it's interesting that in the past few years we've grabbed onto the terms "crowd" and "mob" and applied them as positive characterizations of Internet sociality. Howard Rheingold's *Smart Mobs* in 2003 applied the term to people connected through instantaneous digital communication,[11] and James Surowiecki's *The Wisdom of Crowds*[12] in 2004 pointed to ways that unassociated groups of people can come up with more accurate answers than can individuals. Both books—each excellent—had titles that played upon our negative feelings about groups of people who are sharing space. "You see," both books in effect said, "there's a new positive potential in bringing big groups of strangers together."

Smart mobs and wise crowds represent only two ways that knowledge can be developed on the Internet, often simply by being connected. Let's look at five of the most basic properties of the Internet, starting with the simplest and moving to the more complex. Each of these is giving rise to its own types of networked expertise.

1. The Internet connects lots of people

The first and most obvious fact about the Internet is that it's the biggest crowd anyone has ever seen.

Just as James Surowiecki says, there is a type of expertise that can come from people who are in the same place without being any further organized. *The Wisdom of Crowds* opens with what is now the canonical example. At a county fair in the eighteenth century it was noticed that if you wanted to know how much a particular ox weighs, the average of the total guesses of fair-goers was likely to be closer than the estimate of any particular expert. Surowiecki's book is careful to lay out the precise conditions under which crowds do better than experts—it depends on there being a diversity of opinion, independence, decentralization, and a way to derive a collective decision—but almost as soon as he published it, "the wisdom of crowds" was used to refer to everything from choosing presidents, to the making of best-selling fashions, to voting for your favorite on *American Idol*. That we've so stretched this phrase shows just how excited we were about the new possibilities for social knowledge.

Knowledge always has been social. We clustered experts into think tanks and academic departments because we recognized that they're smarter when together. In the eighteenth century, the great Western thinkers constituted what they called a "Republic of Letters," in which they shared their ideas in correspondence, arguing back and forth at the speed of ponies and sailing ships. Even in ancient Greece, where the idea of knowledge was invented, the most famous thinker reached toward knowledge exclusively through dialogue with others.

But there used to be a natural size to such networks. Few people were admitted to the Republic of Letters, and it really helped to be a leisured white man. University departments are small enclaves. Books and then radio and TV are one-way media, and only a small group of people get to broadcast through them. Within those limits, we constructed a system of knowledge that concentrates expertise in a relative handful of people: If it costs so much to communicate to lots of people, we better give the microphone to those with the most expertise per square inch.

The Internet undoes those constraints. Its massiveness alone gives rise to new possibilities for expertise—that is, for groups of unrelated people to collectively figure something out, or to be a knowledge resource about a topic far too big for any individual expert.

The simplest forms are what Jeff Howe called "crowdsourcing" in a 2006 article in *Wired*.[13] He intended it as a play on "outsourcing," and his examples mainly were of "plugged-in enthusiasts" who will work for much less than traditional employees. But the term was so good that it quickly escaped its creator's tether and now applies to just about any instance in which the mass of the Net lets us do things for little or no cost that otherwise would have been prohibitively expensive.

The examples of crowdsourcing are familiar at this point. When Members of Parliament were found to be routinely taking frivolous deductions, the British newspaper *The Guardian* set up a site where 20,000 people pored through 700,000 expense claims. When technologist Jim Gray and his sailboat went missing on the ocean, Amazon enabled people to scour thousands of satellite images for traces—fruitlessly, it turned out. The Internet enables us to gather and interpret information simply because the Internet is so damn big that you need only a tiny fraction of people to volunteer.

Sometimes the fact that the Internet covers so much physical ground is enough to create crowdsourced expertise. For example, in 2009, the Defense Advanced Research Projects Agency (DARPA)—the R&D branch of the US Department of Defense—decided to celebrate the fortieth anniversary of its creation of ARPANet, the precursor to the Internet. It moored ten 8-foot red weather balloons at accessible locations around the United States and offered a prize for the first accurate report of all of their locations.[14] DARPA wanted to see if social networking sites such as Facebook, MySpace, and Twitter could serve as platforms for quickly gathering intelligence nationwide, a task with implications for national defense. About 4,000 teams entered to compete for the $40,000 prize. Within nine hours, a group at MIT had beaten them all simply by putting up a Web site that promised a substantial cut of the winnings to the first people to report the coordinates of the balloons and smaller cuts to the chains of people who had invited the finders onto the MIT team.[15] What would have been extremely difficult for an individual to know turned out to be a snap for a network.

The popular technology site Engadget pooh-poohed the significance of this contest. "DARPA would have you believe that it's the brilliance of modern-day social networks that led an MIT-based team to win its red balloon challenge this weekend" when it was just that MIT offered to split the money.[16] That objection misses the point: Without the network, the offer of money would have gone nowhere.

Indeed, some of the most powerful ways to crowdsource expertise involve paying people. Amazon's Mechanical Turk, launched in 2005, enables vast numbers of people to work on small, distributed tasks for a small amount of money per transaction. (It's named after an eighteenth-century chess-playing "machine" that beat almost all comers, including Napoleon and Ben Franklin, by concealing a human chess expert within it.) Businesses have used Mechanical Turk to get thousands of online images labeled, find duplications in yellow-page listings, and rate the relevancy of a search engine's results. As my colleague Jonathan Zittrain points out, there is a potential for abuse, such as someone using Mechanical Turk to engage the crowd in matching photos, never letting on that they're being employed by an oppressive government to identify people at protest rallies.[17] But whether Mechanical Turk is being used for good or evil, it is a type of networked expertise that previously would have been prohibitively expensive. Before the network made crowds so available, labeling millions of photos would have required hiring hundreds of professional cataloguers. The professionals at, for example, the digital image vendor Corbis do a more precise job, because they are trained and experienced. But the crowd does a good enough job, at a fraction of the cost. Further, the crowd can scale up to handle massive numbers of photos in a short period if necessary. And it is indeed arguable that crowd-based cataloging (or "tagging") can produce systems of labels that more closely match how users think about matters.[18]

Most of the examples of crowds networking themselves into expertise do not involve money. Far more common are examples like that documented by Calvin Trillin in a light-hearted *New Yorker* article about food-enthusiasts' attempts to track down Peter Chang, a Chinese chef who moves unpredictably from strip mall to strip mall.[19] This informal network of "foodies" found each other through sites such as ChowHound.com, announcing sightings via email and blogs. The primary motivation for this networked collaboration was low on the Maslow scale: Chef Chang is a hell of a cook. But the overall process is so common now that it's practically invisible to us: We rely upon work done by people who don't know one another, spread out across a global network, who have answered a question, gathered data, refined results, contributed to a web of blogs, or even built an encyclopedia. And at the end of the day, the contributors to this networked expertise split a prize, whether it's money, reputation, or "crispy eggplant cut like French fries and salt-fried with scallion greens, a hint of cumin, and hot pepper."

2. The Internet has many different types of people in it

John Davis was a chemist in Bloomington, Illinois, who didn't know a lot about oil. But it turned out that he knew enough about cement to solve a problem that had the oil experts stumped.

In 2007, the Oil Spill Recovery Institute, a nonprofit created by Congress in response to the 1989 *Exxon Valdez* oil disaster,[20] offered $20,000 to the first person who could figure out how to get the oil from that spill out of the bottom of the ocean where it had been sitting for eighteen years. Simply pumping it up did not work because when it got to the surface, the Alaskan air solidified the mix of oil and water, making it impossible to pump off the barges.[21] But Davis knew that cement wouldn't harden so long as you keep vibrating it. Perhaps if the oil were kept stirred up on the barge, it wouldn't harden either.[22] Problem solved. Davis spent part of his prize money to fly himself to the Alaskan site and has offered to work for the Institute for free on other projects.

Davis's wisdom was not that of the crowd. But get a crowd large enough, and you'll find experts like Davis—so long as the network contains many different sorts of people who know many different sorts of things, and so long as it has mechanisms by which the experts can be located. This is exactly the nature of the Net: huge and incredibly diverse. It enables a type of expertise just about impossible to actualize before the Internet existed.

Contests are one way to separate the individual experts from the crowd. It might be a $1,500 first-place prize for the undergraduate who writes the best paper about long-term pavement performance in a contest the Federal Highway Administration has run since 1988.[23] It might be the $500,000 the Department of Transportation has kicked in to the X-Prize for innovation in renewable fuels for the aviation industry.[24] Or it might be the million dollars Prize4Life is offering through InnoCentive for finding a cure for ALS (Lou Gehrig's disease). In each of these cases, the network of experts has value only because that network contains many different types of people.

InnoCentive, a company that spun out of Eli Lilly in 2000, has become the leading contest-based expertise broker, typically offering $10,000 to $100,000 for solutions to problems posed by clients such as Procter & Gamble, NASA, Novartis, and the Rockefeller Foundation. Individuals from anywhere, with any degree of training, can submit solutions. For example, Ed Melcarek, who was earning a living installing HVAC systems, won $25,000 for figuring out a novel way of getting fluoride into toothpaste tubes for Colgate-Palmolive: Ground the tube and give a positive electric charge to the fluoride power.[25]

As Karim Lakhani, a professor at Harvard Business School, has pointed out, such contests occurred well before the advent of the Net. In 1714, the British Parliament offered £20,000 to the first person who could come up with a way of determining longitude at sea.[26] John Harrison was finally granted the money in 1773, at the age of seventy-nine, after a literal lifetime of work on the project.[27] But before the Internet, it pretty much took an act of Parliament to spread word of a contest beyond the bounds of the experts in the field. Now, the Internet's looseness with information has enabled contests to become a normal way in which knotty problems can be solved. For example, you might post to your social networking page news about a contest in your field, but that posting will be seen by people outside your field. Even if you lead such a shuttered life that you only have friends within your narrow range of expertise, they undoubtedly straddle some lines: In a world in which everyone is six degrees from everyone else, it's at the second degree that things start to get really interesting. As news spreads from person to person, it sprays out across far wider networks. This is vital because, as Lakhani's study of InnoCentive discovered, "the further the problem was from the solvers' expertise, the more likely they were to solve it."[28] In other words, the Net enables expertise to emerge not only because so many people are connected to it (property #1) but also because those people are different from one another in how they think and what they know (property #2).

InnoCentive is hardly the only place this happens. TopCopder has been running contests for software developers since 2001 and typically has over one hundred of them open simultaneously. In 2006 Netflix offered a million dollars to the best improvement greater than 10 percent in the algorithms it uses to recommend films to subscribers; in 2009 the team known as BellKor's Pragmatic Chaos won for improving the recommendation algorithm by 10.09 percent.[29] And the Sunlight Foundation, a nonpartisan organization dedicated to increasing government transparency, ran the "Apps for America" contest in 2009 to encourage people to come up with innovative ways to extract value from the mass of agency data the Obama administration is making public. In 2010, it ran the contest again, with $15,000 going to the winner.

But contests are not the only way that nuggets of expertise can be found in the vast pebble field of the crowd. *New York Times* technology columnist David Pogue occasionally throws out a question to the over 1 million people following him on Twitter. In January 2009, to demonstrate the power of Twitter to an audience of 1,000 at a Las Vegas event, he famously asked his followers for a cure for hiccoughs. In less than fifteen seconds, hundreds of answers started rolling in, some serious, many not.[30] But he frequently asks more significant questions as well. For example, in January 2011, Pogue used Twitter to research Verizon's iPhone coverage in New York City by asking his followers to tweet dead spots.[31] He has also polled his followers for far more specific expertise: Anyone know how to save tweets to a text file? Any ideas for managing the ecological consequences of disposable gadgets? What were some of the old computers designed for use in kitchens? Now, Pogue can get his questions answered so quickly and richly because he has 1.3 million Twitter followers, but there have been sites on the Web for over a decade where people who are not famous can pose a question and receive answers from anyone in the crowd. Some medical-advice sites flag the answers from authenticated health care providers, while others rely upon either the crowd thumbs-upping or -downing answers or on the reputation the answerer has earned at the site. Quora, a site that began building a following in 2011, creates a sophisticated social network and reputation system to evaluate the probable worth of the answers volunteered there. In all these cases, the network comes up with answers not for money or to win a contest but because answering questions is a rich social activity that has a wide variety of rewards. The Net's social nature in this case drives it as an information network.

My metaphor for this type of networked expertise as the finding of gems in a vast pebble field is not exactly right. Something else is happening. John Davis was not an expert in cleaning up oil spills until someone asked a question to which his own off-topic expertise suggested a particularly good answer. That probably would not have happened if there were no Internet and if the Oil Spill Recovery Institute had decided instead to post its challenge in its printed newsletter. Only because the question broke out of its tight circle of experts did John Davis become an expert in moving freezing-cold oil. The Net's inability to fence in information—its proneness to "information spills"—let the right pieces connect, so a new idea was born. In cases like these, the Net, as a place that connects lots of people who are different from one another, is not only finding expertise but also generating it.

3. The Internet is like most oatmeal: sticky and lumpy

When a large food company wanted to know how to extend the shelf life of its mashed potatoes, they contacted YourEncore, a network of retired Procter & Gamble employees. The company engaged two retired experts in food microbiology who visited the processing plant, reformulated the product, and made recommendations for changes in production, sanitation, and quality control.

While architecturally the Net connects all nodes equally, socially it consists of billions of sub-networks. YourEncore is one of these: a social network of people joined by membership, mailing lists, and other network tools. Without these sub-networks, the Net would be just a large, flat resource. Instead, clusters naturally form, and it is these lumps of people, pages, and tools that bring the Net most of its value as a place for information, communication, and sociality.

Another network of people and resources, CompanyCommand.com, has an entry requirement even more exclusive than YourEncore's: You have to be a West Point graduate. All graduates of this institution are given command of a company on the ground, but because they had no way of sharing their experiences and learning from one another, a handful of the graduates decided to build a network so they could talk among themselves. About five years later, the US Army embraced the idea fully, creating the Center for the Advancement of Leader Development and Organizational Learning (CALDOL) in 2005 on the West Point campus. What began as a mailing list is now an extensive Web site with discussion areas, multimedia educational materials, and the initial elements of a social networking site. Participants in this large but exclusive network ask practical questions about commanding units, pose challenging problems for group discussion, and talk about the personal issues common among soldiers away from home and sometimes in harm's way.

This particular network is markedly different from the real-world social network in which the participants are embedded. As soldiers walk around West Point's grounds, the markings on their shoulders make manifest exactly where they are in the hierarchy. But when participating in the CALDOL online expertise network, those markings are purposefully kept invisible; it's considered bad social form to pull rank. What counts in the online network is the quality of one's participation, not the marks on one's sleeves. As a result, knowledge and expertise are shared and developed with fewer artificial impediments than one often finds in corporate and educational settings.

Social networks such as CompanyCommand and YourEncore add value to the Internet porridge because they comprise people with expertise that can be brought to bear on some particular class of problems. But, as is typical of the Net, nothing is that simple. Even the lumps can have lumps. That's the case with Expert Labs, the White House–inspired network of members of the American Association for the Advancement of Science, the leading association of scientists in the United States. Expert Labs not only turns the 127,000 members of the AAAS into an expert network, it also opens the network up to anyone who wants to jump in with an idea or suggestion, for the same reasons that InnoCentive doesn't demand that its entrants jump over a credentials hurdle. The professional scientists in the group provide skilled debunking and fact-checking, but the breakthrough ideas are at least as likely to come from people outside the traditional areas of expertise, as Karim Lakhani's research into InnoCentive showed. Expert Labs' first project was a service called ThinkUp that captures the back-and-forth conversations on your social networks at sites such as Facebook and Twitter so that you can slice and dice them, looking for trends, clusters of ideas, and insights. Such networks almost necessarily will include people outside a given scientist's specialty.

Expert networks need lumps—networked conversations and social relationships among the diversely talented people (property #2) in the Internet crowd (property #1)—because expertise multiplies when it exists between people.

4. The Internet is cumulative

A man is turned back at the Canadian border because the customs agent Googles him and discovers something unsavory. A job applicant is all but hired when someone in the Human Resources department finds inappropriate Facebook photos. Political candidates routinely seek to embarrass their opponents by unearthing some dirt on the Web—information that the opponent had proudly posted back in the day when it seemed more funny than dirty. The Net retains everything we post to it—often out of context and sometimes against our will.

But it's not just the dumb things that live on after we've wised up. There are reasons to think that the persistence of smart postings creates a new type of networked expertise. Indeed, you can see expertise accumulating just about in real time. When a new release of a computer operating system comes out, there will be lots of questions and not enough answers. Within days or weeks, however, it is likely that when something goes wrong with the software, you'll be able to get an instant answer by searching for the text of the error message thrown onto your screen. As those answers accumulate, the Net gets more expert about the operating system. These days you're better off waiting a couple of months before installing a new release of software, not because there will be a software fix but because by then the network will have gotten sufficiently expert in the workarounds.

The Net becomes more of an expert not just from the content people create for it but also from the links they—we—draw among the pieces. Linking curates the Net. Yet links are content, too. Indeed, one important type of expertise is being able to run the maze of links. The accumulation of links makes the accumulation of content on the Net ever more usable (because it can be found) and valuable (because a context grows around each piece of content).

The richness of the overall web of content and links enables a healthy mix of types of content-based networked expertise. Some of these collections are carefully vetted, such as medical sites put together by authorized health professionals. Some are process-based, providing functionality that invites people to contribute what they know and to evaluate what others have claimed to know. Some simply enable people to curate content developed elsewhere, linking to articles and debating their merits. And every flavor in between.

Experts, represented and embodied by their individual contributions, are joined by the Net into a networked expertise that can be queried and learned from, that comes up with surprising answers, and that grows faster than anyone can keep up with, because of the Net's sometimes problematic insistence on retaining everything we put online.

5. *The Internet scales indefinitely*

Telephones work great one-to-one, they work okay with five people in a conversation, and they don't work at all one-to-a-million, million-to-one, or million-to-a-million. Television works wonderfully at one-to-100-million but until very recently has not been feasible one-to-one, and still isn't good at ten-to-ten multiway video conferencing. But the Internet works at every scale, and at many scales it allows for an unprecedented back and forth.

Consider Twitter. Twitter works perfectly if you and your five friends are following one another's tweets. It also works if you're Ashton Kutcher and you have millions of people following your every bon mot. It works if you have 100 followers, 10,000 followers, or 6 followers. Twitter works differently at every scale: If you have 6 followers, Twitter is an intimate communication tool, but if you have 1,000,000, it's a broadcast medium. The only other medium that scales this way is paper. And even when the Net is being used for a 1-to-3,000,000 relationship, it enables a type of interactivity no other medium permits. Ashton Kutcher's followers may not be able to communicate with Kutcher, but they can communicate with one another about Kutcher. Conversations may not always scale vertically on the Net, but they do scale horizontally.

With a mere 400,000 employees, IBM's network is not nearly as large as Twitter's, yet the "jams" the company began in 2001 illustrate the benefits that come from the easy way the Net enables many different types of groups to form. In a jam, all employees are invited to address a strategic question of the sort formerly only pondered by the company's biggest brains and highest-paid consultants, using everything from email to wikis to collaborative projects. In 2003, the IBM jam updated the company's core-values statement. In 2006, an "innovation jam" surfaced some big new ideas, five of which—including Smart HealthCare Payment Systems, Intelligent Utility Networks, and the Integrated Mass Transit Information System—have become core to IBM's "smarter planet" initiative. The sessions tend to be creative, unstructured, and heedless of the corporate hierarchy. Jams have been so successful that companies around the world have adopted them, including Nokia, Eli Lilly, and an urban sustainability group that holds a continuing HabitatJam. Then, of course, startups and Web 2.0 companies began holding "jellies," which are like jams but bring together multiple smaller companies.[32]

Because the Net lets us form expert networks of just about any size and configuration, from twosomes to crowds to massively multiplayer games, the expertise of networks need not be equal to the expertise of its smartest member—or even cumulative. The complex, multiway interactions the Net enables means that *networks of experts can be smarter than the sum of their participants.*

For example, BellKor's Pragmatic Chaos was able to win the Netflix prize because the Internet not only made it feasible to assemble experts from around the world but also made it possible for those experts to collaborate. While crowdsourcing can aggregate information—people in every neighborhood of New York City can report on what their local groceries are charging for diapers—networked experts who are talking with one another can build on what they know. We see this all the time on topical mailing lists. Whether the subject is stock picks, knitting, or rocket science (NASA has a social network called SpaceBook), these networks of experts can quickly become the first place received ideas are challenged and the first place new ideas are floated.

That's why the life of the mind is quickly migrating to these networks of experts.

In May 1967, twelve scholars studying the difficult German philosopher Martin Heidegger gathered at Pennsylvania State University and founded the Heidegger Circle. It met once a year, carefully admitting new members by majority vote, until 1998 when it opened up to all who were interested. In the summer of 2005, the group went online, and in 2008 forums were set up.[33] Even though the forums are not all that active by Web standards, they nevertheless contain more people than the old Circle did, because now all that is required for membership is a payment of $35 ($15 for students and unemployed academics). Judging from how people sign their messages, it seems that most of the participants are professional academics, and the discussions are often about topics of peculiar interest to scholars: What did Heidegger mean by the "transcedens" in a particularly obscure paragraph in one of his early works? Was a particular lecture ever published in German? The forum serves—and expresses—the needs of an inner circle of scholars. Because it's online and available 24/7, it enables this self-selected group of eighty-eight to get answers and explore topics just not conceivable in the old 1/365 days of the Heidegger Circle.

Meanwhile, a professor in Costa Rica set up a Facebook page for people interested in Heidegger.[34] It has 1,400 subscribers and is lively with discussion. Clearly not everyone there is an expert, and not everyone focuses full-time on Heidegger. But the discussion is animated, continuous, and unconstrained. It spills out into links to articles that have their own discussion threads. For example, one member links to an article in the *Chronicle of Higher Education*[35] and writes: "Thought you might find this interesting. . . . A new book bashing Heidegger for his Nazi period. . . . Even more interesting is the almost 100 comments defending Heidegger and/or ripping the article's author." Among those hundred comments are deep reflections, studied fact-based arguments, juvenile baitings, and links to yet more discussions and debates. The Heidegger Circle of the 1970s was a closed circle. The Heidegger network of the Web is always connecting to the next new site or idea.

The old Circle had advantages. It was an enclave where you could get productive work done by being with those who shared your basic assumptions. The information exchanged was quite likely to be accurate, the ideas well-considered. And it was an honor to be admitted. But the wall around the club insulated the members from criticism and from outsider points of view that could have helped them. The new network is lively and sprawling, but it contains people who know much less than they think, and can grind important topics down into moist dust.

We don't have to choose between them. Both have value. The Circle is a lump of qualified, sober experts. The Facebook page is a big, throbbing lump of people who want to talk about Heidegger for whatever reason. The two together form a loosely connected network of people who care about Heidegger. The participants collectively know more, they find answers faster, their curiosity is more stimulated, they are made aware of more facets of their topic, and they are involved in more discussions about those facets. The multi-way nature of the Net enables smart experts to be smarter than ever, although it's clear that the Net can also enable us to go down wrong paths with ever more certainty.

Nevertheless, this is a significant and palpable change. Those who suffered from Smartest Guy in the Room syndrome are learning that the rules have changed. When an expert network is functioning at its best, the smartest person in the room is the room itself.

Expertise That Looks Like a Network

The Challenger Commission remains a model for approaching a particular type of problem: Use the power and prestige of the office of the presidency to pull together an elite team of the very best in their fields, fund them heavily, relieve them of some of their ordinary duties so they can devote sufficient time to the project, and give them all the resources a Presidential Commission needs.

But it is not a model that scales. The various forms of networked expertise we have just looked at, each reflecting an essential property of the Net, are enabling new models of expertise to emerge—even within quite traditional sectors.

"Mitre's role is providing expertise to our government clients," says Michal Cenkl, director of Innovation and Technology.[36] To do so, the MITRE Corporation, with a Defense Department pedigree going back to its founding in 1958, certainly could take the Presidential Commission route, bringing together a tight circle of highly competent and highly credentialed experts to produce a final report. Instead, explains Cenkl, "[t]he products we're delivering to our sponsors, in some cases, have evolved from formal written publications—MITRE tech reports— to formats that are more timely and interactive, such as email exchanges, in-person briefings, presentations and discussions, which are much harder to codify and capture." Why? For the sake of speed, but also because the materials that are "harder to codify" represent an interactive dialogue, have a shared context at a specific point in time, and are characterized by a richness of ideas, information, and knowledge. Codifying them would in fact lose some of that context and richness.

"True enough, we have internal experts we can draw on," says Cenkl, "but we've also realized that we need to change the way we present ourselves. It's not necessarily that the MITRE person is the smartest person in the room. We've decided that the model needs to evolve so that we become the brokers of expertise. Our value is that we understand the government's problems really, really well and we can bring the entire community to bear."

Jean Tatalias, MITRE's director of Knowledge Management, says: "We have an environment in which everyone can publish,"[37] instead of designating people as experts and rationing their time. MITRE's internal search engine lets users search for nondesignated experts who nevertheless have been contributing to discussions. "You can see what forums are discussing the topic and what forum to put your question out to," she says. "No one is anointed as the only credentialed expert in a field."

In fact, "[t]he forums don't necessarily come to consensus. There's a desire to use all the expertise available but not a pressure to drive to the right answer for all circumstances," adds Les Holtzblatt, chief Work Practices architect.[38] Why? Because networks of interacting experts are smarter than the accumulation of individual expert opinions whether we're using a simple mailing list or a more highly structured knowledge management community. This is a fundamental distinction not only in the way expertise is derived but also in its nature: MITRE, which is in the expertise business, finds it often delivers more value to its clients when it involves them in a network of experts who have differing opinions.

Of course, experts have always profited by being in communication with one another. But now that expertise is embedded in—and enabled by—a digital network, it is shedding properties of the old medium and taking on properties of the new one:

Expertise was topic-based. Books focus on specific topics because they have to fit between covers. So, in a book-based world, knowledge looks like something that divides into masterable domains. On the Net, topics don't divide up neatly. They connect messily. While people of course still develop deep expertise, the networking of those experts better reflects the overall truth that topic boundaries are often the result of the boundaries of paper.

Expertise's value was the certainty of its conclusions. Books get to speak once. After they're published, it's expensive for the authors to change their minds. So, books try to nail things down. But because the multitude of people on the Internet are different in their interests and abilities, a network of experts is of many minds about just about everything. The value of a network of experts can be in opening things up, not simply coming to unshakable conclusions.

Expertise was often opaque. While experts' reports usually tell us how their conclusions were derived, and typically include supporting data, we don't expect to be able to go back very far in the experts' thinking: The reports and their included data have been our stopping points. Networked expertise puts in links to sources—and even to contradictions—as a matter of course. A simple search is likely to turn up contextualizing information about the expert and about the information the expert is relying on.

Expertise was one-way. Books are the original form of broadcasting, a one-to-many medium: The reader can write a big red "No!!!" in the margins, but the author will never know about it. The Net, on the other hand, is multi-way. Any expert who thinks she will talk and we will simply listen has underestimated the Net. We will comment on her site, and if she doesn't permit comments, we will angrily note that fact on blogs, on Twitter, on Facebook. This multi-way interactivity can make a network of experts more creative, and more responsive to the multitude of ideas and opinions in the world. It can, of course, also create and propagate misinterpretations of the expert's ideas.

Experts were a special class. Relatively few people get to publish books. To do so, you have to pass through editorial filters. Because those filters have generally done a good job, getting a book published both required and bestowed credentials. It might, for example, have gotten you tenure. On the Net, we find expertise emerging from contexts that at first seem unrelated. The person who figures out how to clean up an oil spill may be a cement expert. On the Net, everyone is potentially an expert in something—it all depends on the questions being asked.

Expertise preferred to speak in a single voice. Books have authors and editors who ensure the content is self-consistent. Even anthologies have editors who ensure that there is appropriate consistency among the contributors, at least in terms of content and tone. Experts, too, have been self-consistent; they get embarrassed if caught contradicting themselves. Networked expertise is more like a raucous market of ideas, knowledge, and authority.

This transition from expertise modeled on books to expertise modeled on networks is uncomfortable, especially now as we live through the messy transition. We know the value of traditional expertise. We can see a new type emerging that offers different values. From credentialed to uncredentialed. From certitude to ambivalence. From consistency to plenitude. From the opacity conferred by authority to a constant demand for transparency. From contained and knowable to linked and unmasterable.

Most of all, expertise is moving from being a property of individual experts to becoming a property of the Net. Not all networks raise the level of connective expertise, of course. Some networks are in fact dumber—and more insistent about their dumb views—than its smartest members are. It all depends on the network's particular constellation of the two basic Net properties: The Net is *connective,* and it connects pieces that are *different* from one another. Connectedness is nothing new for experts, although the scope, scale, and transparency of Net connectedness certainly matter. It's the connectedness of a trillion differences that is really shaking up expertise. We paid the big-time experts all that money because we expected them to take the differences buzzing around our brains like gnats and blow them away once and for all. But networked expertise draws its strength from those differences in connection. Indeed, on the Net, the measure of one's strength as an expert often is not that you have the final word on some topic but that you have the first word. And from that first word—whether it's on a blog post, a tweet, or a sheet of old-fashioned white paper—spin out a million gnats of difference, buzzing across the linked world, unsettled and unsettling.

For networked knowledge necessarily contains differences.

5

A Marketplace of Echoes?

W<small>E ARE A BUNDLE OF CONTRADICTIONS THESE DAYS.</small>

On the one hand, we think it's important to have our beliefs challenged, preferably at a fundamental level. *On the other hand,* when the Internet shows us pages and posts that challenge our most basic beliefs, we complain that it's full of people who believe all sorts of crazy things.

On the one hand, we want there to be more serendipity so people won't stay cocooned within their comfort zones. *On the other hand,* just about everyone complains that the Internet is too distracting—too filled with serendipity.

On the one hand, we celebrate the fact that now there are many voices available to us, not just those of the traditional media. *On the other hand,* we complain that all these uncredentialed, unreliable people get megaphones as big as those handed to scholars and trained journalists.

We have these contradictions because the Net's riot of ideas is forcing us to face a tension in our strategy of knowledge that the old medium of knowledge papered over. We thought that knowledge thrives in a lively "marketplace of ideas" because the constraints of paper-based knowledge kept most of the competing ideas outside our local market. Now that we can see just how diverse and divergent the ideas around us are—because Internet filters generally do not actually remove material, but only bring preferred material closer—we find ourselves tremendously confused about the value of this new diversity.

For example, try reading (or, if you are of a certain age, rereading) David Halberstam's award-winning classic *The Best and the Brightest,* published in 1972 as the United States was still dropping tens of thousands of tons of bombs on Vietnam.[1] Halberstam attempts to explain how the Kennedy White House, so full of superbly educated, dedicated men, could have failed so badly in Vietnam. The book's world is populated by household names now known in few households: McGeorge Bundy, George Ball, Chester Bowles . . . the events it discusses are distant, recalled most often as an analogy to our worst current mistake. But Halberstam's question remains deeply unsettling: How did the best and the brightest get us into the Hell of Vietnam? If these men, so well educated and so worldly, erred so badly, how can we trust the advice of lesser men?

Victor Navasky, in his 1972 review of the book, quoted Halberstam's answer: "[T]hey had, for all their brilliance and hubris and sense of themselves, been unwilling to look and learn from the past." Navasky, however, thought the problem was more institutional than individual. If the "good guys" in Halberstam's account had been put into positions of power, perhaps they too would "have been swallowed by the war machine," as Navasky put it.[2]

But the modern reader returning to a land that seems as distant as Oz is amazed that another problem isn't front and center in any discussion of what went wrong in the White House. The best and the brightest are indeed brilliant and well-intentioned. They are hard working. They are patriots. But when you read the book today, you are immediately struck by their sameness. Male. East coast. White. Early middle-aged. Prep school followed by an Ivy League school. Not 100 percent homogenous, but close enough that if they were the board of directors of a corporation today, there would be lawsuits. Had the inner circle included experienced soldiers, old State Department field hands, or maybe even a Vietnamese or two, the White House policy-makers might not have gone so disastrously wrong about the problems they were facing.

Everyone is for diversity these days. Since the days of *The Best and the Brightest,* even our standard stock photos have changed. If a corporate brochure shows more than three employees working together, at least one of them will be a woman and at least one will be African American. While it is perfectly reasonable to criticize a department for not being diverse enough, it takes extraordinary circumstances to make the sentence "I think this group is too diverse" acceptable or even sensible. Diversity as a goal has become the default, in part as a matter of simple fairness, but in part because our culture has long accepted that a diversity of beliefs leads us to better, stronger, more grounded ideas.

Then we look out at the diversity of ideas on the Internet. At AboveTopSecret.com, the forums are filled with people debating whether an airplane really hit the Pentagon on 9/11. Search Google for who killed JFK and, as of this writing, the top results—WhoKilled-JFK.net—puts scare quotes around the phrase "The 'conclusions' of the Warren Commission." If you want to go back not years or decades, search the Web for information about who wrote Shakespeare's plays. After that, you can begin to look for the truly crazy stuff—the people who deny that there are atoms, that the universe is big, that germs cause disease, that two plus two equals four. And all that is just in the English-speaking world. There is so much radical disagreement that we have to rule out the worst of it as beliefs held by the insane—or by cultures that we uncomfortably give up on judging because otherwise we'd have to judge them as insane.

It seems we love diversity until we see what it actually looks like.

Scoping Diversity

When I spoke with Beth Noveck in February 2010, she had a job on hold at the White House, a six-week-old baby, and 2,267 unread email messages. That's what happens when you're on maternity leave from a position as director of President Obama's Open Government Initiative. Until she resigned in January 2011, Noveck was dedicated to opening government processes to all citizens, not just for citizen awareness but for active collaboration. And yet, even as a fierce supporter of open government, Noveck worries about the negative effects of diversity.

Noveck is all for creating networks of experts that look beyond the usual credentials. She gives an example: "If I want to figure out how to get benefits to veterans faster, maybe I should be talking to the people who are on the front lines of the organizations giving out those benefits."[3] She adds, "Do we think about expertise as experiential or academic? Books or context-centered? The person driving a truck every day or the logistics expert from IT [the information technology department]? The answer is both, of course. Now the technology lets you find experienced people as easily as credentialed ones." This opens up the range of opinions and ideas. Plus, Noveck says, it "taps the enthusiasm of someone who is totally self-taught."

So, yay for diversity. But Noveck was in a high-pressure environment with a four-year window for getting things done. She is thus quite blunt about the point at which diversity becomes a problem, not part of the solution. For example, there are federal regulations governing the more than one thousand Federal Advisory Committees (FACs) to ensure that they include a diversity of opinions. Noveck is all for diversity when it makes sense, but as the process gets closer to making actual decisions, it is less and less helpful: If an administration needs help deciding what to do about, say, endangered species, "I don't know that having creationists and evolutionists on the same committee is the best way to get advice."

This is not a hypothetical for Noveck. When she arrived at the White House and quickly put up a public site to solicit proposals for an official open government policy, thousands of people contributed their ideas and discussed the ideas of others, almost always seriously and sometimes in great depth. Yet, the "tag cloud" told a different story. Tags are labels that participants can apply to comments so that they can be more easily sorted and found, and a tag cloud displays those labels in a font size that reflects their frequency. Among the largest tags: "UFO." The "birthers" (people who believe that Barack Obama was not born in the United States) were also out in force, as evidenced by the use of the tag "usurper." Their inclusion in a discussion of open government policy is an example of diversity taken to the point of diversion. Few of us would object to shutting the site to those who want to talk about something completely off-topic, but Noveck had a particular problem in this case. Because it's a government site, removing irrelevant posts would certainly have been claimed by some to be an act of political censorship. So, she had to leave the irrelevant comments up. Fortunately, the community moderation features meant that users could flag posts as off-topic while still leaving them available.

There are two overall points to take from her experience.

First, there is a right degree of diversity. Too little diversity and you end up thinking it's a great idea to invade Vietnam. Too much diversity and you have citizens haranguing you about Hawaiian birth certificates when you're trying to come up with standardized formats for open government data.

Second, what counts as the right degree of diversity is highly context dependent. If you are holding a forum on the legitimacy of the Obama presidency, you want to make sure you have multiple sides represented. If you are organizing a rally to protest the legitimacy of the Obama presidency, you don't want to invite to the planning meetings people who are dead set against you politically.

This is quite a general problem, not just one faced by policy-makers. When Ford was deciding whether to drop its Mercury line of cars, it didn't invite in people who think that all cars should be immediately dumped into the ocean to form new reefs. But if Ford relied only on the same old folks with the same old perspectives and stock options, its decisions would likely be as realistic as those of JFK's best and brightest. The same is true in every field from education to science, although there's variability in the amount of diversity that's helpful before it becomes merely disruptive.

The question, then, is how do you scope diversity appropriately?

1. Not all diversity is equal

Scott Page, in his book *The Difference,* presents evidence to support his bold claim that on or off the Net, "diversity trumps ability."[4] His explanation of this phenomenon is actually quite straightforward:

The best problem solvers tend to be similar; therefore, a collection of the best problem solvers performs little better than any one of them individually. A collection of random, but intelligent, problem solvers tends to be diverse. This diversity allows them to be collectively better.[5]

While "the homogeneous collection may just as well contain only a single person,"[6] Page lists four conditions that have to hold if diversity is to be a better strategy. First, the problem has to be difficult enough that no single problem solver always comes up with the right answer; otherwise, you just need that one brilliant problem solver. Second, the individuals in the group have to be smart relative to the problem; if it's a calculus problem, a diverse group of people who don't know calculus is not going to outperform a single calculus expert. Third, the people in the group have to be able to provide incremental improvements on proposed solutions. Fourth, the group has to be large enough and has to be drawn from a large, genuinely diverse pool. Given these four conditions, it's better to have a diverse group than a group that consists only of the very best minds—diversity trumps the best and the brightest.[7]

But what type of diversity? Getting a group of people who are diverse in their shoe sizes wouldn't help much. Nor does it help to diversify solely by ethnic or racial identity, even though that's how many businesses try to achieve diversity. This is the land mine that Supreme Court Justice Sonia Sotomayor stepped on (or was the hand-grenade lobbed at her, if you prefer) during her confirmation hearings. In various speeches between 1994 and 2003, Sotomayor said that "a wise Latina woman" might come to better conclusions than a white male judge. According to her detractors, she was asserting the superiority of racial or ethnic identity. According to her supporters, she was referring to "the richness of her experiences."[8] Diversity of experience might help to open one's mind to unexpected ideas and to increase one's sympathy for a wider range of people—traits presumably relevant to the sagacity of the nine-person committee to which Sotomayor was applying. But mere diversity of ethnicity is not: Imagine if Sotomayor were a Hispanic whose life experiences were identical to those of the other sitting judges. The racial, gender, and ethnic diversity on which businesses so often focus turns out to be irrelevant except when the individuals have experiential diversity because of those factors.

The Difference presents research showing that the sort of diversity that makes a group smarter than its smartest individuals is a diversity of *perspectives* and *heuristics*. Perspectives are the maps we give to ourselves to represent the lay of the land. For example, if the issue at hand were how to manage comments at an open government site, one perspective might organize the comments by topic, another might look at how they map across the political spectrum, yet another might cluster them by their emotional tone; each of these maps would lead to different approaches to managing the comments. Heuristics, on the other hand, are the tools we bring to bear on problems. For example, a heuristic might be that contentious discussions calm down when there are moderators, or that reputation systems (users giving a thumbs up or down to comments, perhaps) can marginalize the irrelevancies. According to *The Difference*'s analysis, a wise Latina woman will make the Court more diverse in useful ways only if her experiences have given her a different way of looking at the world (perspectives) and taught her different techniques (heuristics) for dealing with issues.

2. Have just enough in common

Mae Tyme is an anti-pornography feminist. Annie Sprinkle is a "prostitute/porn star turned artist/sexologist."[9] Unsurprisingly, they hold very different views about pornography. Those differences are absolute on this topic. Mae Tyme is the pseudonym of a lesbian separatist who believes that "all women that do pornography are either terribly misinformed, or have been enslaved." Tyme would be kicked out of her community for talking to a "porn star," according to Sprinkle.[10] Yet, in 2000, Sprinkle and Tyme sat down for a conversation, which Sprinkle then posted on her Web site. It is a frank interchange. Sprinkle argues that porn can be "liberating" and goes so far as to recommend that people make their own porn in order to learn more about sex. Tyme says that pornographers engage in a form of child slavery, and even compares Sprinkle to a Holocaust denier. Despite the starkness of their differences, the conversation ends with each saying why the other is "a good teacher for me," and hugging.[11]

This seems to be evidence of the power of diversity: Two people completely opposed on an issue nevertheless have a respectful conversation in which each learns from the other. Tyme learns that female porn stars can find the profession "liberating and profitable"—and that the women are generally paid more than the men. Sprinkle learns that opposition to pornography does not have to stem from a fear of erotic experience. But mainly they both learn to respect each other despite their deep differences on the issue of pornography.

And yet, we could draw two other lessons that are not nearly as encouraging about the possibility of people coming together in respectful conversation.

First, the difference between Sprinkle and Mae is overshadowed by what they have in common. Yes, they are diametrically opposed on the issue of pornography. But they are both committed feminists who have organized their lives around that core value. They also share a vocabulary: the words "porn," "radical lesbian," "dyke," and "patriarchy" have precise enough meanings for them that they can use them without first defining them. Many other conversations would have been derailed by wrangling over these issues: "How dare you call me a porn star! I'm a sexually honest actor in the adult entertainment business!" In addition, from the conversation we can safely assume that she and Sprinkle are both women, speak English, are over forty years old, and are able to talk frankly about sex and sexuality. Take these away, and the conversation would have become far more difficult, or even impossible.

Second, the very last thing Mae Tyme says is: "A conversation like this is possible when each of us has freedom of expression and no one is required to change. I don't expect you to become anti-porn, and you don't expect me to become pro-porn." Had the discussants insisted that there be a winner and a loser—or if they were on a committee charged with setting a policy for a business or the government—the conversation would have been far more difficult. Tyme and Sprinkle entered the conversation with an implicit understanding that the goal was simply to explore an alternative view held by someone they already like and respect.

Unfortunately, the sorts of conversations we often have in mind when we think about civility are precisely ones in which people who strongly disagree need to come to agreement about some practical matter of policy. But diversity works best when there are shared goals, as Scott Page points out. That's why Beth Noveck doesn't want to invite climate change deniers into the discussion of practical steps to slow down climate change. It's not because she's close-minded. It's because the goal of that discussion is to slow the warming of the earth. Among people who share that goal, there is plenty of room for diversity of perspectives and heuristics. A diverse group of people who share a goal are likely to be more effective than a homogeneous group of people. Communities of knowers need walls around them. Those walls used to be like those of a fortress. These days, they tend to be usefully semipermeable. But they are walls nonetheless, and serve the good purpose that walls of every sort serve: permitting a group with enough in common to get something done by keeping out disruptive diversity.

Too much commonality leads to groupthink. Too little commonality leads to wheel-spinning or committees that compromise toward mediocrity. The trick is to have just enough diversity. And just enough in this case is usually measured in scoops smaller than we had assumed.

3. Mix well by hand

A cookbook with a soufflé recipe that tells you to add "just enough milk" would be worse than useless. The same is true of a book that tells you to add "just enough diversity." How much is enough? We know that it's considerably less than we thought, but the problem is that there is no set amount of diversity that is just enough.

That's why The WELL has moderators. Founded in 1975 by the generational icon Stewart Brand, with Larry Brilliant, The WELL has been one of the longest-running conversations on the Net. Its origins are in the hippie culture of which Brand is an avatar—the name stands for The Whole Earth 'Lectronic Link, a reference to Brand's *Whole Earth Catalog*—but the 4,000 current members seem to reflect more of an earnest coffee-shop culture than the shirtless non-linearity of Haight-Ashbury. Jon Lebkowsky, who has been on The WELL since

1987, says that the site's success was not accidental. "They were successful in building the community by seeding it originally with people who were great conversationalists," waiving the fees for the people they wanted involved. Lebkowsky adds, "They also invited the Grateful Dead crowd, and a lot of journalists."[12]

After the seeding comes the gardening. Every "conference"—a topic of conversation, some of which have been running since The WELL's inception—has two "hosts," or moderators. Moderators are allowed to participate, but they are there primarily to keep the conversation just diverse enough. If someone becomes uncivil beyond the norms for the group, the moderator may step in. On occasion, people are banned from the discussion for a cooling-off period—the conversational equivalent of a time-out. If the conversation steers off course, the moderator may remind people what they are there to discuss. Howard Rheingold, one of the founding parents of online discussion and a denizen of The WELL since 1985, urges community forums to have moderators. Even the mere presence of moderators—even if they never moderate a single posting—is enough to keep out the trolls, he says.[13]

Moderation does not have to occur through designated moderators. Sometimes there is simply too much traffic to make that feasible. Community moderation frequently does the trick—as at Beth Noveck's OpenGov site, where the group of people actually interested in open government policy moderated the "birthers" into their own corner. But moderation almost always requires dedicated people for the same reason that the right amount of diversity cannot be specified beyond "just enough." Human conversations reflect every dimension of human individuality, sociality, and culture. There isn't a single rule that we wouldn't want broken at some point. No foul language in a parenting discussion? Okay, but how about when the discussion turns to what to do about a potty-mouthed child and it becomes important to know exactly what words are being used? No hate speech in a political discussion? Okay, but how about when the discussion is about the language being used at some more hateful discussion site? English-only? Fine, until someone asks for help translating the angry Spanish remarks of a customer who refused to speak English. Stay on topic in a scientific discussion? Great, until a desperate parent apologetically interrupts with a question about a sick child. We need moderators because conversations cannot be entirely rule-based. How much diversity is "just enough" depends on the people, the topic, the purpose, the social bonds—on everything.

4. Fork it over

Unfortunately, we have had to remove this feature, at least temporarily, because a few readers were flooding the site with inappropriate material.

Thanks and apologies to the thousands of people who logged on in the right spirit.[14]

That's what greets you when you browse over to the *Los Angeles Times*'s "wikitorial" experiment. Back in mid-2005, when the success of Wikipedia was whipping up Wiki Fever, it seemed that even the most contentious issue could be resolved if people were allowed to work together on a webpage devoted to it. So, when the *LA Times* ran an editorial called "War and Consequences" asking the Bush administration to clarify its plan for withdrawing troops from Iraq, it followed up with an invitation:

How do you like the above editorial? A lot? Thanks! Not so much? Do you see fatuous reasoning, a selective reading of the facts, a lack of poetry? Well, what are you going to do about it? You could send us an e-mail (or even write us a letter, if you can find a stamp). But today you have a new option: Rewrite the editorial yourself, using a Web page known as a "wiki," at latimes.com/wiki.[15]

Four days later, the *LA Times* ran the following wiki-bituary:

The *Los Angeles Times* has canceled a novel Internet feature that allowed readers to rewrite an editorial on the newspaper's website, after some users sabotaged the site with foul language and pornographic images.

The newspaper launched the experimental "wikitorial" Friday and killed it early Sunday after an unknown user or users posted explicit photos.[16]

Within two days, the original editorial had been edited 150 times. At one point it was turned into an editorial critical of the role that the *LA Times* had played in the run-up to the war. Comparisons to the Philippine-American War were inserted by some people and removed by others.[17] And then, of course, there were the disgusting images repeatedly posted by vandals.

Jeff Jarvis, an important voice for openness in the debate about the future of journalism, blogged that "[a] wikitorial is bound to turn into a tug-of-war" and suggested that an alternative wiki page be set up for those who disagreed with the editorial. The founder of Wikipedia, Jimmy Wales, responded that he had already done so, creating a "counterpoint" wiki on the *Los Angeles Times* site for those who differed from the newspaper's view.[18] "I'm not sure the LA Times wants me setting policy for their site," wrote Wales, "but it is a wiki after all, and what was there made no sense."[19]

No sense at all. Wikis try to get everyone on the same page, quite literally. But when a diversity of passionate opinion is inevitable—no editorial has ever been powerful enough to dispel all contrary ideas—a wiki is exactly the wrong idea. A better idea is to enable the discussion to fork. Forking is a familiar tactic. For example, when on a mailing list a few people start flinging emails back and forth on a topic of marginal interest, someone might sensibly suggest that they take it off the group and carry the conversation through private email. At The WELL, under those circumstances, a moderator will suggest creating a separate topic thread. In the real world, discussions fork when people break off from a group to go talk among themselves.

The Internet is perfect for forking. It's got infinite space where the divergent conversation can continue out of earshot of everyone except those who choose to hear it: If it's a mailing list, you don't get the forked discussion's emails unless you ask to participate. Forking enables a group to find its own level of diversity.

Then again, there are those who see forking as the Net's fatal flaw. . . .

Into the Echo Chamber

Up until this point, I've maintained that because the Internet shows us how much there is to know and how deeply we disagree about everything, our old strategy of knowing by reducing what there is to be known—knowledge that is shaped like the data-information-knowledge-wisdom pyramid—is badly adapted to the new ecology. Instead, we are adopting strategies that take advantage of our new medium's near-infinite capacity. As a result, our basic idea of what knowledge looks like and how it works has been changing.

Yet, three of the four tactics for dealing with diversity we just looked at recommend reductive tactics: Get just enough diversity, use a moderator to keep the diversity from getting too great, and fork discussions when they become too diverse. What's going on? Has the accessibility of this super-abundance of ideas and knowledge changed nothing? In fact, has the super-abundance of knowledge made us more narrow-minded? Fork "birthers"—or enthusiastic supporters of President Obama—into their own discussion, and they're likely to close themselves to external criticism and egg one another on, rather than be opened up by a good diverse conversation. This happens not just in goal-directed policy discussions. Everywhere on the Net, people are forking themselves into groups of like-minded people because it is fun to engage with people who share our enthusiasms, but also because we can't get our shared work done if we constantly have to argue about first principles.

Groups that fork themselves so tightly that they include only people who agree with them are called "echo chambers." If people are living in echo chambers on the Internet, then it doesn't matter how many differences, disagreements, and points of view are present outside each chamber. If we're holing ourselves up with people who think exactly the way we do, then knowledge is hiding from diversity, excluding more differences than ever before.

If the Net is creating more echo chambers, the biggest loser will be democracy, for the citizenry will be polarized and thus be less able to come to agreement, and to compromise when it cannot. This is perhaps the greatest concern expressed by Cass Sunstein, a constitutional scholar and currently the administrator of the White House Office of Information and Regulatory Affairs. Sunstein, who is the most-cited living legal scholar in the United States,[20] has written a couple of books on the topic. In *Republic.com,* published in 2001, he argues that when people get to choose what they see, they will tend toward that which is familiar, comfortable, and reinforcing of their existing beliefs, a tendency others call "homophily."[21] Sunstein shows the distressing power of homophily by pointing out that "[i]f you take the ten most highly rated television programs for whites, and then take the ten most highly rated programs for African-Americans, you will find little overlap between them. Indeed, seven of the ten programs most highly rated by African-Americans rank as the very least popular programs for whites."[22]

"Similar divisions can be found on the Internet," he adds.[23] He lists sites that are explicitly designed for African Americans, for young women, for young men, and so on. He also cites research that he and a colleague did that found that of sixty randomly chosen political sites, only 15 percent put in links to sites of their opponents. "[M]any people are mostly hearing more and louder echoes of their own voices,"[24] because the Internet so increases the range of choices that citizens can find narrowly focused groups that precisely mirror their point of view.

The news is even worse than that, Sunstein fears. Studies have shown that when people speak only with those with whom they agree, they not only become more convinced of their own views, they tend to adopt more extreme versions of those views.[25] This group polarization happens for two reasons, Sunstein says. First, the members of the group have a smaller pool of views from which to drink. Second, because people "want to be perceived favorably by other group members," they will often adjust their views toward the dominant position. "In countless studies, exactly this pattern is observed."[26]

The picture Sunstein paints is scary for those who care about democracy and disappointing to those who had hoped that the Internet would move us toward the traditional ideal of a knowledgeable person: open-minded, fact-oriented, and eager to explore other perspectives. Sunstein's studies of group polarization specifically looked at offline interactions. So, we need to know: Is the Net in fact closing our minds and moving us toward more extreme views?

Sunstein is convinced: "Group polarization is unquestionably occurring on the Internet."[27] His evidence is that "it seems plain that the Internet is serving, for many, as a breeding ground for extremism,"[28] pointing to "cybercascades" in which a belief rapidly gains many believers because it is being passed around the Net as true. Plus, "[a] number of studies have shown group polarization in Internet-like settings."[29] But, how big a problem is it on the Net? As Sunstein acknowledges in an edition of *Republic.com* published a year later, "To know whether this is a serious problem we need much more information."[30] Does it happen a lot? A little? All the time? Compared to what? How often? How much? Does the Internet's diversity of sources ever depolarize some groups? If so, why those and not others? Perhaps, as Clay Shirky has suggested, Sunstein has it "exactly backwards": Perhaps "political discourse is coarsening" not because people are walling themselves into echo chambers but "precisely because people are constantly exposed to other points of view."[31]

Indeed, some recent evidence suggests the polarization may not be as extreme as Sunstein thinks.[32] Economists Matthew Gentzkow and Jesse Shapiro published a paper in 2010 that looked not at which sites link to which but what sites individual users actually visit as they spend time on the Net.[33] This study's results seem to be the opposite of what Sunstein's "group polarization" idea would lead us to expect: "Visitors of extreme conservative sites such as rushlimbaugh.com and glennbeck.com are more likely than a typical online news reader to have visited nytimes.com. Visitors of extreme liberal sites such as thinkprogress.org and moveon.org are more likely than a typical online news reader to have visited foxnews.com."[34] That is, those visiting the most obvious examples of partisan echo chambers are also more likely than most people to visit sites on the other side of the political divide.

So, is the Net reducing our shared experience, leading to group polarization, and thus hurting democracy by making us narrower knowers than ever? The Gentzkow-Shapiro study suggests not, but it's just one study, and it is subject to dispute. For example, Ethan Zuckerman, my colleague at the Berkman Center, took a careful look at it and drew exactly the opposite conclusions.[35] He points out that the study finds that Net users are more insular than users of just about all the old media. Indeed, if we were simply to look around the Net, using our own experience as a guide—the opposite of a careful methodology, granted—many of us would, like Cass Sunstein, conclude that people do seem to be more polarized and more uncivil than ever. If you want to attract attention on the Internet, talking in extremes seems to be an effective tactic.

We are not yet close to having a solid answer to Sunstein's question. Yet, it's worth noting that it always seems to be "those other folks" who are being made stupid by the Net. Most of us feel, as we're Googling around, that the Net is making us smarter—better informed (with more answers at our literal fingertips), better able to explore a topic, better able to find the points of view that explain and contextualize that which we don't yet understand.

Not Nicholas Carr. He thinks the Net is making all of us stupider, including himself, but more or less for the opposite reason that Sunstein worries about. Carr notes at the beginning of his wonderfully titled book *The Shallows* that he realized in 2007 that his own cognitive processes were changing because of the Net, and not for the better. "I missed my old brain," he writes.[36] For Carr, the cause is not the presence of echo chambers but their rough opposite: The linking, blinking, twittering diversity of the Net is making us dumb. The Web is reshaping our physical brains, Carr contends, "weakening our capacities for the kind of 'deep processing' that underpins 'mindful knowledge acquisition, inductive analysis, critical thinking, imagination, and reflection.'"[37] He cites studies of the brain and of behavior to prove that the Internet is getting us not only to think differently but to think worse.

Carr's picture accords with what many of us have sensed: These days we seem to be more easily distracted, we have less patience for long books, we want to jump over the boring parts to get to the "meat," we have difficulty remembering how we got to where we are on the Web. At the same time, the studies Carr cites do not accord with the sense many of us have that we are now smarter than we were because the only limit on how quickly we can get answers is our typing speed, and because our curiosity has to travel only the length of a finger flick to be satisfied, and then be aroused again.

We all know that some of the places where we are smartest work only because they have properties of echo chambers: The clamor of disagreeing voices is muffled or silenced. Knowledge has always needed communities to flourish. Communities need walls so that they can let in the right amount of diversity, even if too frequently they err on the side of homogeneity. But now the Net has made community walls semipermeable. The transparency of the Net lets outsiders look in and insiders look out. And you may be exchanging ideas in a community that Cass Sunstein would call an echo chamber, but you got there by passing through the daily chaotic roil of ideas on the Net. Our old echo chambers were like quiet libraries in quiet communities. Our new echo chambers—knowledge communities—are on the busiest street in the world and there are no windows thick enough to cut out all the noise.

So, is the Net making us stupider or smarter? The Net is new, the research is relatively scant, and the Net is rapidly evolving. The answers may well—in fact, probably do—vary by the usual variables in such studies: economic level, education, gender, politics, interests, geography, culture, and so on. The concept of "echo chambers" is itself slippery. And then there is the difficulty of measuring any quality as culturally determined as "smartness." As Carr writes, "The Net is making us smarter . . . only if we define intelligence by the Net's own standards."[38] The answer to the question "Is the Net making us smarter or stupider?" is going to be settled not by thinking through the problem but by living through it.

Yet, there is a sense in which it does not matter. Whether or not the Web tends to make us more insular, we know that human beings have a tendency toward homophily; we prefer to be with people who are like us. All the participants in this debate agree that excessive homophily is a bad thing. All the participants agree that we should be bending our efforts to work against our homophilitic tendencies. And no participants—not Cass Sunstein, not Nicholas Carr—are suggesting that we roll the Net back up and throw it away as a bad idea. So, why so many years of debate and with such passion?

Because something else is at stake.

Unsettled Discourses

Al Gore published *The Assault on Reason*[39] in 2007 in the middle of George Bush's second term,[40] so it's understandable that he felt some despair. "Why do reason, logic, and truth seem to play a sharply diminished role in the way America now makes important decisions?" he asks on the first page. After many chapters convincingly making the case that governance has become unmoored from fact and argument, Gore talks about the Internet as a "source of great hope for the future vitality of democracy."[41]

> [T]he Internet is perhaps the greatest source of hope for reestablishing an open communications environment in which the conversation of democracy can flourish. . . . The ideas that individuals contribute are dealt with, in the main, according to the rules of a meritocracy of ideas. It is the most interactive medium in history and the one with the greatest potential for connecting individuals to one another and to a universe of knowledge.[42]

Even though for Al Gore the Internet is democracy's hope and for Cass Sunstein it is democracy's danger, the two men agree on an underlying premise: The way to truth and knowledge is through reasonable and open encounters among those who disagree.

This idea has been with us for a long time. Socrates thought so. The picture of reasonable people sitting together, talking over their differences in a respectful, honest way is the image on the Enlightenment's own Hallmark Card. Jürgen Habermas, the influential German philosopher, marks the beginning of the "public sphere" with the rise of public spaces such as coffee shops where those conversations could happen. Al Gore is hopeful because he sees the Internet as an expansion of this reasonable public sphere. Sunstein is worried because he thinks we're retreating into semi-private spheres. But we have all—almost all—thought that the way forward through our manifest differences has been to be open to contrary ideas, and to talk about them reasonably.

There's a set of presuppositions behind this belief about the power of reason, the purpose of conversation, and the relation of knowledge and the world. Even if we leave those presuppositions alone, we don't have to spend much time on the Net to come to the sad conclusion that simply as a matter of fact, we're not ever going to learn to talk together reasonably and come to single conclusions. We are going to disagree about everything. That's the fact of diversity with which we now have to deal.

What do we do about it? When it comes to climate change, Al Gore's general strategy has been to say that those who disagree with him are not within the community of Reasonable People: The deniers are wrong on the facts, they don't believe in science, and thus they have no place in the Coffee Shop of Reason. Evolutionary scientists often treat creationists the same way. And the other side uses exactly the same tactic: Al Gore is a close-minded hysteric with whom there is no arguing; evolutionists are Godless rationalists who won't acknowledge the reality of the Divine so there's not enough common ground to even begin a discussion. The only place we have the sort of rational discussions Gore, Sunstein, and Socrates value so highly is within an echo chamber—a room in which people agree thoroughly enough that they can disagree reasonably.

For example, in May 2010, Republicans in Congress launched AmericaSpeakingOut.com where people could post ideas that the Republican Party congressional leadership said they would read. The Republicans hailed this as a "revolutionary" democratization of the governing process. Then they watched in horror as, within the first couple of days, the ideas posted included the repealing of Section II of the 1964 Civil Rights Act because it is (as the suggester "explained") "UNCONSTITUTIONAL, PROGRESSIVE and HITLER."[43] Another suggested putting all American Muslims under surveillance. Another suggested a tax hike. The Republicans allowed the Muslim surveillance and tax hike suggestions to remain on the site, but asserted that they are not going to consider acting on those ideas. "The key is to remember that we are focusing on . . . settled principles," said Representative Peter Roskam of Illinois. Those settled principles formed the outer edges of the permissible discussion. Stray beyond them and your ideas would be ignored or, in extreme cases, removed. This means America SpeakingOut.com is an echo chamber. But that's a requirement if Republicans are going to make progress discussing the issues in the ways that define them as Republicans.

It's important to be clear about this. We still need as much difference and diversity within the conversation as we can manage. We still need to continuously learn how to manage to include more diversity. We need to be on guard against the psychological tricks echo chambers play, convincing us that our beliefs are "obviously" true and nudging us toward more extreme versions of them. But it's also fine for the Republicans to have an online "coffee shop" where they can discuss their ideas. That conversation needs far more agreement than diversity.

The Internet is showing us that our old ideal of a Coffee Shop of Reason exists only within a city with millions of other coffee shops that look wrong, wrong-headed, or totally unreasonable. But that fragmentation is exactly what the Age of Reason thought we could overcome. We now have pretty good evidence that we cannot. That evidence is the Net itself.

———

For a couple of thousand years, Western philosophers have debated whether human reason is sufficient for understanding our world. But the critics of reason generally have been lonely. In the past fifty years, these voices have become a chorus that in some quarters has become the loudest singers around. They got labeled "postmodernists," which has stuck no matter how much they've objected.

When I was a graduate student in philosophy, the postmodernists hadn't yet had much impact on North American soil. I wrote my dissertation on Martin Heidegger, a contemporary German philosopher who was considered by mainstream philosophers to be purposefully obscure and prone to making "everything you know is wrong" statements simply to appear deeper than everyone else. But I, and the academics in my circle, thought the obscurity of Heidegger's writings was due to the profundity of their challenge to the fundamental assumptions of Western philosophy. Then, in 1978, postmodernism became an unexpectedly central topic at the annual Heidegger Circle meeting. Much of the talk, especially in the hallways, centered on writings by Jacques Derrida, who maintained that Heidegger's radical philosophical project had not gone far enough. As a new member getting to hang out with the scholars whose work had guided me, I was struck by the fact that the sorts of things they were saying about Derrida were precisely the sorts of things non-Heideggerians said about Heidegger: He was out to shock, incoherent, purposefully vague, an intellectual charlatan. We were, in short, having a classic echo chamber moment, refusing to take seriously claims that challenged our own. The irony is, of course, that the rest of the world would consider Derrida and Heidegger to be overwhelmingly alike in their ideas.

Over the years, I have struggled with the works of this new generation of thinkers. Postmodernist writings tend to be remarkably dense, either—depending on who you talk with—because they are attempting to undo deep, basic assumptions embedded in language itself or because they are using a fog of language to hide the emptiness of their ideas. Obviously, no brief introduction can claim to be adequate, especially since there are so many differences among them. Fortunately, we need only a few crucial ideas from them to help us understand the world of difference the Net exposes to its every visitor.

All knowledge and experience is an interpretation. The world is one way and not others—the stone you stubbed your toe on is really there, and polio vaccine works quite reliably—but our experience of the world is always from a point of view, looking at some features and not others.

Interpretations are social. Interpretation always occurs within a culture, a language, a history, and a human project we care about. The tree is lumber to the woodcutter, a place to climb to the child, and an object of worship to the Druid. This inevitably adds human elements of uncertainty and incompleteness.

There is no privileged position. There are always many ways to interpret anything, and none can claim to be the single best way out of its context. Some postmodernists talk about this in terms of denying that there are "privileged" positions, intentionally invoking not only the Einsteinian sense (all motion is relative) but also, pointedly, the socioeconomic sense (the elite should not get to marginalize the ideas of the rest).

Interpretations occur in discourses. You can't make sense of something outside of a context. Even something as simple as a car's turn signal can only be understood within a context that includes cars, the basics of physics, the unpredictable intentions of other drivers, the restrictions of law, and the way left and right travels with one's body. Ludwig Wittgenstein talked about this in terms of "language games," by which he meant not something you do for fun but, rather, the way our words and actions are guided by implicit rules and expectations. Postmodernists have many different words for these contexts, but we'll use the term "discourses."

Within a discourse, some interpretations are privileged. If you are within the discourse of science, fact-based evidence carries special weight, and emotions do not. On the other hand, if you propose marriage by compiling tables of data as if you were within a scientific discourse, you either are making a bad joke or are seriously disturbed. Discourses are themselves social constructions—they are ways people within a culture put ideas together. They are not themselves part of nature, and they change throughout history.

There is much, much more to postmodernism. But these five ideas are crucial and, in various forms, have been resisted and debated for decades now. That was, of course, before the Internet showed us that the postmodernists were right. Freed of the privileged position accorded them by the limitations of paper-based media, the old, centralized authorities are losing their purchase. On the Net we see just about every possible interpretation. When these interpretations contend across discourses—or across cultures, or socioeconomic groups, or any other group that has its own norms and values—there is little possibility of resolving the differences and sometimes even of understanding them.

And yet we cannot afford to let the best and the brightest stick to their own discourse, because when we do, hundreds of thousands of people can die in a needless war. Put people on a network and they might form the sort of echo chamber that Cass Sunstein worries about, making themselves more certain, more extreme, and more dangerous as they reconfirm their old opinions with white papers and backslaps. For those who have no interest in intellectual rigor, or who lack curiosity (which, by the way, characterizes each of us for at least part of every day), the Net may well be an environment that degrades knowledge. We need to be concerned about all this. And we need much more research to ascertain the actual risk and actual damage. But the network also offers the possibility of connecting across boundaries, forming expert networks that are smarter than their smartest participants. The network can make us smarter *if we want to be smarter*.

But we can't leave it there. The question of difference drives us to hold paradoxical positions—the Net is an echo chamber, the Net distracts us with all its diversity—for the same reason that we have been so resistant to postmodernist ideas. We worry, understandably, that without a privileged position we will be lost in a swirl of contradictory ideas.

So, let's take a look at how networked knowledge puts ideas together, especially as compared with the old, bookish ways by which we used to work our way toward the truth. Do the differences and distractions on the Net result in an inevitable frittering away of ideas? Is the hyperlinking of ideas an admission of failure, a new way forward, or a bit of both?

6

Long Form, Web Form

All men are mortal.
Socrates was a man.
Therefore, Socrates was mortal.

THAT HAS BECOME A STANDARD EXAMPLE of how to know something.[1] If the first and second lines of this argument are true, the conclusion can be known with a certainty that even God's mighty hand could not shake.

But of course to know the world, we need much longer chains of argument, for the world is a complex whole. We should be able to start at A and reason our way to Z, in careful, measured steps. This *long-form* argument is what we've taken to be human reasoning at its best.

So, what if the Internet is shortening our attention spans? Suppose we can no longer even get from A to B without being distracted by a catch-the-monkey ad or a link to the latest gossip? How we are ever going to think the thoughts that step us well beyond what we already know?

If we're going to worry about losing long-form thinking, we should be quite clear about what it looks like. One of the greatest of long-form works was published in 1859. Darwin's *On the Origin of Species* is a single, magnificent argument, spread out across fifteen chapters. Here it is in summary, with chapters marked:

[Introduction] We've had various ideas about the origins of species. Let's take a new look. [1] Farmers breed new varieties of domesticated animals by selecting parents with the traits they want. [2] There's also

lots of variation in wild animals. [3] The variations that help wild animals to survive better enable them to have more offspring, passing
along those variations. This is natural selection, much like the artificial
selection done by farmers. [4] Natural selection operates in small steps,
and explains why even apparently useless features have developed. [5]
There are some laws about how variations occur.

[6–7] Some may object to this theory. Let me address those objections. [8] Natural selection can also explain the development and inheritance of instinctive behaviors. [9] While it's true that this theory
cannot explain why hybrids are sterile, since sterility is the opposite of
a reproductive advantage, there is another explanation. [10] While we
don't have fossils that show the complete record of evolution, there are
good reasons for that. [11] In fact, the fossil record, understood correctly, supports the gradual evolution of species by natural selection.

[12–14] The variation of species over the face of the earth confirms
that species adapt to their environments. And here are four more
pieces of evidence that support my argument.

Finally, [15] "As this whole volume is one long argument, it may be
convenient to the reader to have the leading facts and inferences
briefly recapitulated."

A successful long-form work lays out the argument in careful steps,
it deals with objections, it provides support, it concludes. *On the Origin
of Species* is as brilliant a piece of literature as it is a foundational work
of science.

Even so, it suffers from the weaknesses of all long-form arguments.
For example, in Chapter Four, Darwin writes: "He who rejects these
views on the nature of the geological record, will rightly reject my
whole theory." Part of the genius of the book is Darwin's awareness of
his reader's desire to jump off the bus before it pulls into the station.
So, Darwin spends a full six out of fifteen chapters addressing objections he imagines his readers may have. How could natural selection
explain the loss of features, such as the nonfunctioning eyes of some
bats? Why don't fossils show transitions between species? Darwin
presents the objections of his colleagues, as well as anticipating criti-
cisms not yet raised. Brilliant.

But brilliant within the constraints of the medium available to him. If you're writing a book, you have to have a conversation with yourself about possible objections because books are a disconnected, nonconversational, one-way medium. We have had to resort to this sort of play-acting not because that's how thought should work but because books fix thoughts on paper. We've had to build a long sequence of thoughts, one leading to another, because books put one page after another. Long-form thinking looks the way it does because books shaped it that way. And because books have been knowledge's medium, we have thought that that's how knowledge *should* be shaped.

For example, Robert Darnton, a renowned historian of books and director of the Harvard University Library, makes this point in a 1999 essay of his reprinted in *The Case for Books*:

> Any historian who has done long stints of research knows the frustration over his or her inability to communicate the fathomlessness of the archives and the bottomlessness of the past. If only my reader could have a look inside this book . . . all the letters in it, not just the lines from the letter I am quoting. If only I could follow that trail in my text just as I pursued it through the dossiers, when I felt free to take detours. . . . If only I could show how themes criss-cross outside my narrative and extend far beyond boundaries of my book. . . . [I]f instead of using an argument to close a case, [books] could open up new ways of making sense of the evidence, new possibilities of apprehending the raw material embedded in the story, a new consciousness of the complexities involved in construing the past.[2]

In fact, in this essay, Darnton described how a book might incorporate these capabilities: "[S]tructure it in layers arranged like a pyramid."[3] At the top would be the "concise account." Second, there could be "expanded versions of different aspects of the argument." Third, there could be documentation to support the top two layers. Fourth, include "selections from previous scholarship and discussions of them." Fifth would be teaching tools. The sixth layer would aggregate reader commentary and exchanges.

This sounds like an interesting proposal for structuring a Web site or an e-book, but Darnton in 1999 was proposing a new type of physical book. "The computer screen would be used for sampling and searching," he writes, so that readers could specify what they want included in a book printed specifically for them, so that "concentrated, long-term reading would take place by means of the conventional codex"—that is, in a printed, bound book.[4] Why? Because the physical book "has proven to be a marvelous machine"—convenient, comfortable, universally accessible, "a delight to the eye," a "pleasure to hold in the hand."[5]

Darnton is a fascinating mix of a connoisseur of physical books and an advocate for progressive library policies.[6] He's been outspoken in the push for open access to digital works, and has been a leader in the struggle to get the needs of readers and libraries accounted for in Google Books. And Darnton is right that physical books are not going away. Neither did live theater, yet it's no longer the dominant cultural form of performed works. Likewise, physical books will no longer be the dominant cultural form of knowledge, if only because the physical book is such a bad fit for the structure of knowledge it's intended to represent and enable. The historian's *cri de coeur* Darnton begins with—"frustration over his or her inability to communicate the fathomlessness of the archives and the bottomlessness of the past"—calls out for a far more fluid, more highly connected, more interactive form that now, well past 1999, we have and that Darnton appreciates.

At last thought has a medium that helps it past the limitations of physical books that brought us to think of long-form thought as the highest and most natural shape knowledge could assume. But what shape does networked knowledge tend toward? Short-form thought? Narrow-form thought? Or, perhaps the idea of shape gets in the way when we're trying to understand knowledge.

Book-Shaped Thought

I am aware that it is at best ironic, and at worst hypocritical, that I have written a long-form book, available only on paper (or on paper's disconnected electronic simulacrum), that is arguing for the strengths of networks over books. My apology is of the unfortunate sort that does not justify the action so much as humiliate the perpetrator. And so: I am sixty years old as I write this, and am of a generation that takes the publication of a book as an achievement—my parents would have been proud. It's also not irrelevant to me that book publishers still pay advances. Beyond these primordial and pathetic motivations—seeking money and Mommy's approval—there are some other factors that mitigate the irony. I'm not saying "Books bad. Net good." The privilege of holding the floor for the length of 70,000 words can allow ideas to develop in useful ways; if this book spends more time discussing networks than books, it's because its author assumes that the case for books is made implicitly by every schoolroom with bookshelves, every paragraph of flap copy, and every public library. Further, for the past fifteen years I've been working in a hybrid mode that is not inappropriate to the transformation we're living through: I have been out on the Web with the ideas in this book since before the book was conceived, and have profited greatly from the online conversations about them. (Thank you blogosphere! Thank you commenters!) Still, not only is the irony/hypocrisy of this book inescapable, it is so familiar in this time of transition that I wish someone would write a boilerplate paragraph that all authors of nonpessimistic books about the Internet could just insert and be done with.

Nicholas Carr's book *The Shallows* escapes the irony because it maintains that long-form books are the crucial and distinctive way civilization develops ideas. If there's any irony at all, it's that Carr's long-form book aims to convince us that the Internet is reshaping our brains so that we can no longer follow long-form arguments—since *The Shallows* is indeed a coherent, 220-page argument.

We could outline it as we did *On the Origin of Species,* chapter by chapter: [1] We all sense that our way of thinking is being changed by the Internet, and we regret it. [2] The brain is remarkably plastic, molding itself to new needs and inputs. [3] Historically, technology has changed how we think. [4] Deep reading and thinking developed historically because of books. [5] The Internet is a new and quite different technology. It is changing how we think. [6] E-books will change how we read and write, from solitary, private concentration to connected, public, hyperlinked frenzy. [7] Science tells us that these changes in behavior and mental abilities are in fact due to the Net's rewiring of our brains. [8] Google, the dominant and emblematic tool for guiding inquiry on the Web, is taking thought down a bad path marked by the "strip-mining" of meaning and a "pinched conception of the human mind."[7] [9] We are not merely offloading memory to our machines. The scientific study of memory reveals that the Web is making us dumber by damaging our long-term memory, impeding the development of conceptual schemas, reducing our ability to pay attention, and, worse, threatening "the depth and the distinctiveness of the culture we all share." [10] Worse still, this is affecting our very souls.

Reducing a rich long-form book to a single paragraph does violence to a work that needs many more pages to be expressed—if Nick Carr could have tweeted his book, he would have. But the reduction does show that his work has a logical form in which parts depend upon the parts before it. *The Shallows* is a good example of a modern long-form work.

But how exactly did it assume that form? Carr's book started as a famously controversial cover story in *The Atlantic* magazine, "Is Google Making Us Stoopid?" Undoubtedly a complex set of considerations then came into play: It's a rich topic that needs further development, it might make a successful book, and so on. But the question that concerns us is not Nicholas Carr's motivations. Rather, it's how the nature of the book *qua* book affected the form of its content. For, all this talk—Carr's, mine, many others'—about how technology shapes thought is far from metaphorical or abstract, as if the Platonic form of the Book molds us to its will through mystical emanations. When you sit down to write a book, the bound pages—the boundness, the pageness—make demands of you. Of course you could fill the pages with scribbles or use them to start a fire, but you sat down to write a book. So, you begin.

Books have beginnings because bound pages have a first page. You could specify that the reader start on page 135 and read outward, or you could just order the pages randomly, but that's not what books want from you. So, you think about what is a starting place your readers will accept.

The pages continue, so you continue. That the pages are bound, and thus are in sequence, demands that your ideas have a sequence. So, you continue not simply by writing words but by finding the continuity among ideas.

You finish. The length is up to you, although there are physical limits on how many pages the shop can bind, as well as upper and lower limits on the page count that your publisher will accept. Even if you slop over into a second volume because your ideas are as rich and complex as Darwin's investigation of barnacles, your book has an end and thus needs an ending. You write a sentence that leaves the reader feeling that you've done the job. You imagine the reader's sigh as she closes the back cover, satisfied that the book has wrapped up the argument, but (we authors hope) regretful that the journey that began on page 1 is over.

The book is published. It is out. It stays constant as the world changes around it.

The physical book's demands have thus had you reinvent long-form writing. The book develops an idea from start to finish, across many—but not too many—pages. It has to contain within its covers everything relevant to that idea because there is no easy way for the reader to access the rest of what she might need. You the author determine the sequence of the ideas. The book's physical finality encourages a finality of thought: You don't finish writing it until you believe you have it done and right.

The physical nature of books thereby enables and encourages long-form thought. "Enables and encourages" because the physical nature of books is not enough to entirely account for it: To a different culture on a different path of thought, bound pages might look like an encouragement to divide thought up into page-sized chunks, as with PowerPoint. There were long-form narratives and investigations before book makers cut scrolls into uniform pages and bound them. We came up with a medium that suited how we were already structuring thought on rectangular surfaces. That medium has remarkable advantages, but it also has characteristics that unintentionally limited and shaped knowledge. Books do not express the nature of knowledge. They express the nature of knowledge committed to paper cut into pages without regard for the edges of ideas, bound together, printed in mass quantities, and distributed, all within boundaries set by an economic system.

To think that knowledge itself is shaped like books is to marvel that a rock fits so well in its hole in the ground.

The Embarrassment of Books

Just as the slippery hold of slotted screwdrivers became obvious only after x-shaped Phillips-head screwdrivers became common, many of the disadvantages of printed, bound books are becoming obvious only now that a medium with a different physics is taking hold. The glorification of the old medium often sounds like the sublimation of an embarrassment about the sudden exposure of that old medium's weaknesses.

Sven Birkerts probably would not agree. In 1992 he wrote a book that was a milestone and is perhaps a classic: *The Gutenberg Elegies*.[8] He wrote this before the Web, when hyperlinking existed only within closed systems that compiled documents as if they were software programs, which of course they were. Electronic books were things on CDs. Reading *The Gutenberg Elegies* now, one remembers that before the Web, electronic communication felt like a diminishment—speech reduced to green dot–based characters—and not like today's overwhelming efflorescence. That makes Birkerts's perspicuity all the more impressive.

So, I went this afternoon to find my copy of his book. I haven't read it in perhaps ten years, and I had only a dim hunch about where it might be—not which shelf but which room. Surprisingly, my first guess was pretty much right: I ran my fingers over the collection of books in the corner shelf in my bedroom—mainly vaguely literary titles, with some detective mysteries and travel guides thrown in—until

I saw the title I was looking for. I pulled it out carefully, because the bedroom corner bookshelf is more a Jenga puzzle than a library space. I carried the book downstairs, feeling a little proud of myself. While we are adoring the tactile pleasures of physical books, let's remember that their physicality makes the misplacing of them a common experience; except for the very orderly among us, the more books you own, the harder it is to find any one of them. We don't count that against books because it is an inevitable consequence of being made of atoms. But now that we have books made of bits that we can find by typing in an approximation of the author's name, the simple act of looking for a physical book can seem like the slipping of a slotted screwdriver.

Because I hadn't come across anything by Birkerts for a while, I Googled "Sven Birkerts." To tell the whole truth, I Googled "sven berkerts." But Google suggested the right spelling, and a split second later I had found what I was looking for: the publication date of his book. I browsed down the Google search results. Ah, he wrote an article in *The Atlantic* in 2009 called "Resisting the Kindle," but which might have been titled "Is Kindle Making Us Stoopid?" to keep it in line with what seems to be the old medium's new theme. In it, Birkerts reiterates one of the most interesting and even beautiful of the arguments in the book he had written seventeen years earlier. In "Resisting the Kindle," Birkerts says that the historical nature of literature is "reinforced by our libraries and bookstores, by the obvious adjacency of certain texts, the fact of which telegraphs the cumulative time-bound nature of the enterprise."[9] Although libraries and bookstores tend to make texts adjacent based on topic or alphabetical order, not historical context, Birkerts's point rings true to me: Books bring forth the past. In the *Elegies*, he writes: "Say what you will about books, they not only mark the backward trail, but they also encode this sense of obstacle, of otherness. . . . Old-style textual research may feel like an unnecessarily slow burrowing, but it is itself an instruction: It confirms that time is a force as implacable as gravity."[10] Beautiful. Indeed, in a library, we have the sense that the past is present, waiting to speak to us, even if we're there to check out the latest Jennifer Aniston romcom DVD. We have the sense that the library shelves go all the way back to the Greeks, to the

Egyptians, to the Hebrews. We don't have that sense on the Net. The Net is a continuous wave front of presence.

Birkerts is such a lovely writer that I enter a bibliophilic reverie. I am in a classic library—I personally envisage the Harvard Law Library where I work, an elegant epitome of the beauty of book culture—where I'm sitting in a leather-bound armchair reading a leather-bound book by one of those ancient writers I've always meant to read. Then I look at the actual book in my hand, a fifteen-year-old paperback of *The Gutenberg Elegies*. The top—the only part left exposed to the air in my bedroom—is dusty. When I open it, the dried glue crinkles and the pages begin to separate from the spine. I thumb through, afraid to spread it wider than the angle of a twig on an autumn branch. The outer margins of the pages look like they have been dipped in weak coffee. The book smells like an item from the past that was forgotten, abandoned. This is not what Birkerts means by books making the past present to us. This actual book's past is present in its decrepitude. Instead of enjoying the frisson of connection to our culture's continuous glory, I have to suppress a sneeze.

We have idealized books. Romanticized them. Some have gone so far as to fetishize them. Our image of them as cultural objects often expresses an odd nostalgia for British reading rooms and dry sherry. But the fact is that most of the books most readers deal with are cheaply made disposable objects. Birkerts's own book is as covered with marketing decals as a NASCAR car, although Birkerts's decals are thoughtful blurbs from esteemed figures. It's not that Birkerts's cover is especially crass or tawdry. Not at all. It is typical. And that is the point: The books we actually read and live with are not the books we imagine. As is characteristic of nostalgia, we remember the glow of an experience and forget the gloom encompassing it.

—

The type of thought that bound, printed, paged, published books encourage has many shadows. For example, Nicholas Carr's book leads us to a conclusion. It is not a simple conclusion such as "Eating white foods makes you fat" or "If you want it, you will have it!" because Carr is a subtle thinker. It's not even "The Internet is making us nothing but stoopid!" because Carr is a far more honest writer than that. His book is aimed at establishing a focused cluster of ideas intended to make us anxious that the Net is irrevocably degrading our thought by altering our brains. Because of this, when he talks about, for example, the way in which books created the experience of thought as inner and private, he ignores other factors. For example, Jean de Joinville's 1309 biography of Louis IX was able to take an early step toward telling the story of an interior life in part because Catholicism was becoming a religion of interiorized penance, which itself was a factor in long-term historical, military, and economic developments. Yet, there is not a word about this in Nicholas Carr's account.

I'm not criticizing Carr for this. On the contrary. His book is not primarily about the development of our interior voice, and his account of the topic is actually quite good. The problem is not with Carr's book. It's the way in which books squeeze ideas onto long, narrow paths that head the reader forward. There are an endless number of influences on developments as fundamental as the interiorizing of thought, and most have nothing to do with the point Carr is making about what we're losing as we make the transition from physical books to the Internet. Long-form writers are out to move their readers from A to the Z that is the author's destination. What is not needed to get readers to that conclusion is left aside. In fact, Carr interrupts his own book with a set of short digressions so he can include ideas that don't fit into his argumentative narrative. He has to position them as interruptions—unusual in books, and structurally awkward—because the physicality of books tends toward sequence, not divergence. Worthy ideas that diverge from the narrative's narrow path appear as distractions. Books often just aren't long enough to enable long-form ideas to uncurl into their natural shape.

Further, we've elevated private thought because of the limitations of writing. The unwritten law of writing physical books has been "One page, one hand." The physics of books generally makes writing them a solo project. So, Nicholas Carr tells us that he moved from Boston to Colorado, downgraded to a "relatively poky DSL connection,"[11] and checked his email infrequently. It was painful, but, he tells us, "[s]ome old, disused neural circuits" sprang "back to life."[12] He moved so that he could be alone, as if thinking in public—what we normally would call "talking"—is a distraction rather than a condition for thinking. Of course, no book could be composed by a truly private person because no person can be truly private. As Carr would acknowledge—for one thing, his book has a page of acknowledgments—the public's contribution to his book is there, if in the shadows. He talked with his wife, he browsed the Internet during his designated hours, he was undoubtedly in communication with his editor, he floated in the sea of publicness (as do all of us) as he composed his initial ideas. Not to mention the publicness of language itself.[13]

I should here confess that my own seemingly erudite comments about Jean de Joinville came directly from an email exchange with Jacob Albert, a summer intern at Harvard's Berkman Center for Internet & Society. Jacob had expressed an interest in the topic of this book, which led to an ongoing discussion. I had never heard of de Joinville when Jacob first mentioned him. The truth is that all I knew about Louis IX was that he very likely came after Louis VIII and before Louis X. Some facts and ideas in Carr's book inevitably came from similar sorts of interchanges. Thought is never private.

Nor should it be. When Carr's initial article came out in *The Atlantic,* there was some wonderful discussion of it among the elite set of thinkers—including Carr—who converse at Edge.org.[14] Danny Hillis, a computing pioneer, agrees that something is making us stupid, but thinks that the "the flood of information" is the culprit. He also points to the role of politics. The writer Kevin Kelly wonders whether Nietzsche's prose "changed from arguments to aphorisms" because he started using a typewriter, as Carr says, or if it was because "Nietzsche was ill and dying." Larry Sanger, co-founder (and then critic) of Wikipedia, agrees that we're becoming less able to string together thoughts, but thinks we should be blaming ourselves, not our technology. The writer Douglas Rushkoff thinks that Carr is correctly noting a change but is getting his values wrong: This is an evolutionary transformation in which the old fish think those new footed youngsters are up to no good. Edge.org also compiles the discussion that spills across other sites, where it continues in loose-edge ways, sometimes with replies from Carr himself, removing the single-point-of-failure that plagues long-form arguments.

Isn't this—a network of thought of any length and form—a better way to know our world?

Public Thinking

If you were to create a map of the influence of bloggers who write about journalism, based on who links to them, Jay Rosen's site, PressThink.org, would be a hub, with many more spidery links going into it than coming out, although next to nothing compared to the large media sites themselves. If you were then to visualize influence by weighing more heavily links from those who are highly read, Rosen's site would inflate under this new projection the way Greenland does on schoolroom maps. But even this new map would hide something important: the shape of the thinking that goes on in and around Rosen's blog.

In one sense, Rosen has been engaged in long-form thinking. Not only does he write blog posts that are famously many times longer than typical, but they cluster around some recurring points. For example, for two years, he has written a series of posts about "what the Internet is doing to the kind of [journalistic] authority once easily constructed by objectivity."[15] Rosen makes it easy to summarize the argument of these posts because at the end of the latest post—he told me he is working on one more in the series—he ran a linked set of descriptions of all six posts:[16]

1. Audience Atomization Overcome: Why the Internet Weakens the Authority of the Press (January 12, 2009)

2. He Said, She Said Journalism: Lame Formula in the Land of the Active User (April 12, 2009)

3. The Quest for Innocence and the Loss of Reality in Political Journalism (February 21, 2010)

4. Clowns to the Left of Me, Jokers to the Right: On the Actual Ideology of the American Press (June 14, 2010)

5. Fixing the Ideology Problem in Our Political Press: A Reply to *The Atlantic*'s Marc Ambinder (June 22, 2010)

6. Objectivity as a Form of Persuasion: A Few Notes for Marcus Brauchli (July 7, 2010)

Together, these constitute a long-form argument. Rosen challenges our belief that good reporting must be objective by giving a different, more plausible reading of the journalistic enterprise. In one of the earlier articles in the series, he defines reporting as "gathering information, talking to people who know, trying to verify and clarify what actually happened and to portray the range of views as they emerge from events."[17] This makes it easier for journalists to accept Rosen's argument in the sixth post that the real value of reporting is not its supposed detachment from all viewpoints but, rather, the fact that the reporter was where the reader was not. The pretense of objectivity can actually get in the way of the reporting itself. Intellectually freed from the assumption that the value of reporting is its objectivity, readers can now explore with Rosen other possibilities for journalism.

The six posts combined contain 110,000 words, which make the series longer than *The Shallows*, and about one and a half times as long as this book. Of those 110,000 words, Rosen wrote only 15,000—about twice the length of this chapter. The rest are comments left by readers. Rosen's posts, however, attract high-quality comments. No "You rock!" or "You suck!" replies. Rather, the comments tend to be well-developed and thoughtful. Rosen told me, "The second most common reaction I have had to my blog (after Why are your posts so long? . . .) is, 'No offense, Jay, but the comments are often better than the post.' It has something to do with the tone the author (that's me) is setting in the original post."[18] He does not review comments before they are posted, and the only ones he removes are outright spam.

There are, of course, advantages and disadvantages to doing long-form thinking in Rosen's public way.

I can think of nine advantages:

First, the argument assumes a natural length. Rosen doesn't have to worry about stretching his ideas from 15,000 words to fill a 60,000-word book container. Nor does he have to worry that the comments are getting too long to include in a single volume, about which he'd have to have a losing argument with his publisher.

Second, the argument is more responsive to the ground it covers. Readers point out topics that the author then realizes he must address.

Third, the work becomes embedded in a loose-edged discussion that more naturally reflects the messy, tangled topology of topics.

Fourth, readers are given fewer reasons to get off the bus midway. When Darwin writes in Chapter Four that "[h]e who rejects these views on the nature of the geological record, will rightly reject my whole theory," he's opening the door and inviting passengers to get off. If Darwin had published Rosen's way, he would have discovered unanticipated objections, and he would have been able to meet at least some of them. This seems like a much more natural rhetorical form than holing-up authors in garrets or in cottages in the wilds of Colorado so that they can imagine the objections readers might have.

Fifth, Rosen's ideas get out to their public far faster than the old write-in-private, publish-when-done model.

Sixth, the ideas more successfully escape the grasp of the author so that they can change the world. The networked nature of

Rosen's writing means that the ideas get passed around easily and may be taken up by people who have never heard of him.

Seventh, readers are more intellectually and emotionally involved because now they can be part of the discussion.

Eighth, the author's authority gets right-sized. Rather than speaking in a Solomonic voice because a publisher has handed him a scepter, Rosen has authority because of what he's saying and because of how others are visibly responding to him. Simply seeing the author engage with readers through comments tells the great percentage of readers who do not leave comments that Rosen recognizes that his words are more tentative than their inscription makes them seem, that his authority goes no further than the worth of his ideas. This changes the source and meaning of authorial authority.

Ninth, it's not just the author who is no longer alone in his wilderness cabin. The readers are also now connected. We can see some of the effects of the writer's words rippling through the culture. We can see that journalists are weighing in on Rosen's site, that other sites are linking back to Rosen, that jayrosen_nyu has 37,000 Twitter followers. We used to have to be told that ideas have effects. Now we can see ideas spreading, like watching on a monitor as dye injected into the bloodstream traces its path.

But there are disadvantages:

For some, the voices of readers may function as noise, as a distraction.

Second, some arguments work better rhetorically if they are presented all at once.

Third, some ideas won't do well commercially if developed in public for free. It's not clear that our assumptions here are correct, though. The fiction writer and activist Cory Doctorow, among others, has succeeded commercially, as well as in the impact of his ideas, by giving away online access to his books even as he sells paper copies.

Fourth, the published book is a traditional token of expertise and achievement.

Fifth, it's harder for us to know what to believe when many voices are audibly in contention.

So, nine advantages to five disadvantages. If only determining the winner were as easy as counting. For example, the fact that webby long-form thinking makes it harder to know whom to believe is not necessarily a disadvantage.[19] But it's not as if long-form and web-form arguments are in contention, and one must be vanquished. There's plenty of room on the Internet for traditional multi-volume long-form works, as well as arguments conducted purely in tweets. At the very moment that Nicholas Carr's work is flying off the shelves of bookstores, Carr can be in the midst of a hot and heavy Web chat fest. Long-form works are discussed on the Net even when they're not available on the Net. This is a both-and, with an extra and-how.

Nevertheless, the Net does not leave long-form argument unaffected. You can certainly write 70,000 words that are hermetically sealed by the logic of their interconnection. You can be Spinoza, who wrote a work of ethics that's structured like a work of deductive geometry. But if that work is going to have any effect, it's going to be put into a network where the discussion around it, and through which readers now come to it, will violate its pristine logic. The stoopidest of the stoopid will misunderstand it entirely and call the author the N or C word, no matter the author's race or gender. Bright but twisted minds will get it just wrong enough to make the author look plausibly like an idiot. Self-interested experts will find it threatening to their position on the professional ladder and will whistle while quietly filing away at the rung it stands on. College students will randomly copy entire pages, claim them as their own, and not understand how they could ever have been caught.

Welcome to the life of knowledge once it has been taken down from its shelf. It is misquoted, degraded, enhanced, incorporated, passed around through a thousand degrees of misunderstanding, and assimilated to the point of invisibility. It was ever so. Now we can see it happening. When the process of knowing occurs online, in our midst, with a comment section and abundant links to other opinions, it's no longer possible to separate knowledge-at-work from knowledge-as-it-is-understood.

So, what is the shape of this new knowledge? What is the opposite of long form? That is the wrong question. Networks of knowledge on the Net have no shape because the Net has no outer edge. Besides, it doesn't stay still long enough.

Shape matters. When knowledge was a pyramid, when it was based on firm foundations shared by all members of the community, when it consisted of contents filtered by reliable authorities, when we knew what was in and what was out, when it had a form and shape, knowledge had an easy authority. The shapelessness of knowledge reflects its reinvigoration, but at the cost of removing the central points of authority around which business, culture, science, and government used to pivot. That change is in itself a serious issue, of course, but it is the result of a change in what traditionally has grounded knowledge: its basic relationship to the world.

From Stopping Point to Temptation

Objectivity has so fallen out of favor in our culture that in 1996 the Society of Professional Journalists' Code of Ethics dropped it as an official value.[20] It's not that journalists decided to be biased and unfair. Rather, objectivity promised something it could not deliver: Reporters would present the situation exactly as it is, independent of any prejudices or individual standpoints. So, many journalists now talk about fairness, accuracy, and balance instead of objectivity. Nevertheless, we haven't really given up on the picture of knowledge that lets objectivity make sense as a concept. It is the same picture that has enabled us to elevate long-form writing—books—to the pinnacle of how humans know the world.

As an example of the problems with objectivity, let's take the coverage of a scripted event by two reputable newspapers that lean the same way politically:

In 2004, Ted Kennedy, the Lion of the Senate, delivered a much-anticipated speech on the second night of the Democratic National Convention, in his home town of Boston.[21] The front-page report in the *Boston Globe* began:

> The second night of the Democratic National Convention featured harsher criticism of the Bush administration, with Senator Edward M. Kennedy accusing the president of making the world a more dangerous place for Americans.[22]

After a paragraph quoting Teresa Heinz, the article notes that Ronald Reagan's son spoke, and Barack Obama "offered a glimpse of . . . what may be the future of their party." The next six paragraphs are devoted to Kennedy's speech, highlighting his stirring call "to take up the cause."

The front-page coverage at the *Washington Post* began without even a mention of Kennedy's speech:

> On the second night of its national convention, the Democratic Party introduced two newcomers to the nation to set the themes that John F. Kerry hopes will help him win the White House in 2004.[23]

When the article got around to mentioning Kennedy's speech, its summary was brief and its take-away was harsh: "Kennedy's effort fell flat in the hall."

The headlines in the two newspapers reflected the deep differences in how they saw the night. The *Boston Globe*'s headline featured tradition and aggressiveness: "Kennedy Leads the Attack: Convention Speakers Rip Bush in Shift of Rhetoric." The *Washington Post*'s headline stressed new voices and unity: "Democrats Focus on Healing Divisions; Addressing Convention, Newcomers Set Themes." There is no overlap between the quotations from Kennedy's speech in the *Post* and in the *Globe*. There is little overlap in what they paraphrase of it. They disagree about whether the speech was inspiring or disastrous. Two major, highly professional newspapers. Both liberal bastions. At the same event. A scripted event. With the scripts literally handed out before the event. And yet, even granting that Kennedy is a Boston hometown favorite, their straightforward reports are close to diametrically opposed.

Many of us who reject objectivity as anything except an "aspiration"—a common idea among journalists—will still find the extent of the difference in the *Post* and *Globe*'s reporting surprising. I did. Our traditional worldview attributes the differences in our reports about that world to human limitations: biases, incomplete information, subjectivity.[24] Objectivity makes the promise to the reader that the report shows the world as it is by getting rid of (or at least minimizing) the individual, subjective elements, providing, as Jay Rosen (echoing the philosopher Thomas Nagel) disparagingly calls it, "the view from nowhere."[25]

Objectivity rests on a metaphysical description of our relation to the world: Real events are experienced by individual minds that strive to create an accurate inner representation, which is then expressed in words presented to others. But objectivity arose as a public value largely as a way of addressing a limitation of paper as a medium for knowledge. For example, America's first magazine, *The American Magazine,* in 1741 promised to "inviolably observe an exact neutrality" because "several colonies have no printing press" and thus "it is difficult to obtain publication for any but a one-sided view of a subject."[26] The current widespread belief that news stories need to be balanced—resulting in what are derided as "he-said-she-said" stories—springs from the same motive: If a single objective account is not possible, then at least give the reader both sides so that the report is complete enough to serve as a stopping point. Objectivity and balance thus address the same limitation that drives long-form arguments: Paper is such an inconveniently disconnected medium that it's important to include everything that the reader needs in order to understand a topic.

As we have lost faith in objectivity (a process that began before the Net arrived), transparency has begun to do much of the work formerly accomplished by it. Transparency comes in at least two flavors. Transparency about the reporter's *standpoint* has been a topic in journalism at least since the "New Journalism" of the 1970s and the "gonzo journalism" practiced by Hunter S. Thompson.[27] For example, Jay Rosen's blog not only takes explicit stands, it has a prominent link to "Q & A about the blog's POV" that lays out his point of view about journalism and tells us that politically he's a "standard Upper West Side Liberal Jewish babyboomer."[28] The ease with which readers can look up information about an author can make their standpoints transparent even if they don't want them to be.

The second type of transparency—transparency of *sources*—is more disruptive of the old system. Paper-based citations are like nails: If you wonder why the author made a particular claim, you can see that it's nailed down by a footnote. Paper-based citations attempt to keep the reader within the article, while providing the address of where the source material resides for the highly motivated researcher. On the Net, hyperlinks are less nails than invitations. Indeed, many of the links are not to source material but to elaborations, contradictions, and opinions that the author may not fully endorse. They beckon the reader out of the article. Links are a visible manifestation of the author giving up any claim to completeness or even sufficiency; links invite the reader to browse the network in which the work is enmeshed, an acknowledgment that thinking is something that we do together. Networked knowledge is thus less a system of stopping points than a web of temptations.

And this points to a second problem with objectivity. The first problem is that humans inevitably understand their world from a particular standpoint. But suppose the problem is not merely human. Suppose the world is itself not as much one-way-and-not-another as we'd thought.

For example, as revolution spread from Tunisia to Egypt at the start of 2011, a controversy arose about how much credit social media such as Facebook and Twitter ought to get. Malcolm Gladwell, the author of *The Tipping Point,* had written a *New Yorker* article in October 2010 arguing that social media are overrated as tools of social change because they enable only "weak ties" among people, instead of the "strong ties" activists need in order to put themselves at risk.[29] When a few months later some media and bloggers credited social media in the Mideast revolutions of 2011, Gladwell posted a 200-word essay asserting that the influence of social media was "the least interesting fact."[30] Gladwell's comments were a corrective to those who carelessly referred to the events as "Facebook revolutions" or "Twitter revolutions" as if they were the sole cause, but he also disputed those who thought social media played a significant role at all. Given Gladwell's standing, and the fact that *The Tipping Point* is about the importance of social networks, his position surprised many. But my point is not that Gladwell is mistaken (although I think he is). It's that even if we do accept that social media played a role of some significance, it's not at all clear what role they played. The more one looks at the question, the clearer it becomes that we don't even have an agreed-upon explanatory framework within which the question might be resolved. And this is true not only of questions touching the Internet. For example, a couple of months after the *New Yorker* ran the original Gladwell piece, it published an article by Louis Menand that similarly wondered how to gauge the social and political effects of books such as Betty Friedan's 1963 *The Feminine Mystique.*[31] We look at social media at work in civil unrest and we wonder how much the media shape us. How does it happen? Does media influence have the same effects on all cultures? On all strata of society? How much of social unrest in general (and in particular countries) comes about as the result of having access to information? How much as the result of communication? Of sociality? If there were no social media, would the revolutions have happened, and, if so, how might they have been different?

The problem with *knowing* the role of social media in the recent Mideast revolutions is that the events themselves are the result of a complex cluster of details that defies predictability and complete under-standing. The same is true for human events overall, which is why we're still arguing about whether the Civil War was fought over the issue of slavery.[32] The world is too intertwingly, to use a word coined by network visionary Ted Nelson—too complexly interdependent and entangled to be fully comprehensible.[33] The messy web of links that transparency gives rise to reflects that intertwingularity. It should lead us to wonder if one of the problems with objectivity and long-form argument is that they aren't a good match to the structure of the world. Perhaps inter-twingly networks reflect the world more accurately than does an "objec-tive" news report or a walk along a long form's narrow path.

—

We do need stopping points, especially when the issues are less ill defined than "What happened at the Democratic Convention last night?" or "What role did social media play in a recent revolution?" or "How did Betty Friedan's writing change history?" We frequently just need an answer so we can get on with our project. That's why we have sites like WolframAlpha.com. The polymath Stephen Wolfram launched the site in 2009 to provide reliable, accurate answers to ques-tions that can range from the prosaic ("How far away is the moon?") to the whimsical ("How many gallons of milk would it take to fill up the moon?"), as well as far more abstruse scientific and mathematical in-quiries. The sorts of questions it deals with have answers that can be known either by looking them up in a reputable source or by comput-ing them from known information. For this, the project gathers cu-rated supplies of information of many sorts and uses a variety of manual and automated techniques for checking their accuracy.[34] We should trust WolframAlpha's facts and its computations as much as we trusted the information in almanacs and other professionally edited source books, and for exactly the same reasons: They've been through editorial filters, they've been fact-checked, and reputations and busi-nesses are at stake. WolframAlpha applies paper-based techniques to the networked medium, to good purpose, and improves upon them by being up to date and by computing answers in real time.

Even so, WolframAlpha is not a stopping point exactly like the older stopping points of knowledge. If a result on the site surprises you, you can click on the link that WolframAlpha always provides to see its sources. If you think the site made an error, there's a blank box brightly labeled "Give us your feedback" at the bottom of the page. And there are links to related searches you might want to explore ("Volume of moon versus earth?"). Of course, the old almanac listed its sources and may have given you an address to send corrections to, but it was a slow, one-way medium with a dash of communication thrown in. WolframAlpha assumes a far greater level of engagement and assumes that you may want to turn your result into a link itself. So, while WolframAlpha uses the old paper-based techniques of authority, it does so within a network that changes the nature of the authorities presented as stopping points. When the authorities were functionally invisible to us—the editors of the encyclopedia, the authors of the textbooks—it was easy to imagine that the chain of authority simply ended there. Now we can see that the people in the box are not the end of the story. They are in linked networks themselves. The chain of authority has no end. We will accept authorities for many of the old reasons, as well as some new ones, but more than ever we know that authorities are stopping points because we choose to stop with them. Transparency shows us that we could choose to go on.

Just as we still need stopping points, we still need long-form writing. But the same sort of thing is happening to it. Long-form writing is often a better instrument of understanding when it is embedded in a web of ideas, conversations, and arguments, all linked and traversable. The writings of Charles Darwin, Nicholas Carr, and Jay Rosen are more useful, understandable, verifiable, and up to date because of the links that point into them and out from them. The links not only let us easily engage with the works, they show us how the rest of our culture is engaging with them. Long-form writing is by no means unnecessary or "dead." But the fact that it is improved by being placed into the Net's web of connections means it is being dethroned by that web as the single best way to assemble ideas.

This is far from all positive.

We trust WolframAlpha, NYTimes.com, Britannica.com, and the Center for Disease Control site because we know that, as with our old authorities, qualified, credentialed people exercise strong editorial control. But the majority of sites on the Net have no professional editors. We rely on them—if we do—for a wide variety of reasons: because what the site says makes sense to us, because it links to sources, because the author has some traditional credentials, because the site is held in high esteem by others in our circle of trust, because the site has a reputation system like Amazon's or eBay's, or perhaps because we're fooled by the author's brazenness and choice of fonts. We can unknowingly find ourselves in a circle of pointers among sites that don't have a grain of truth among them, each reinforcing the others with scholarly looking footnotes. We can put ourselves into echo chambers that repeat lies until they seem obvious. As Cass Sunstein says, there are "information cascades" of false and harmful ideas on the Net that not only gain velocity from the ease with which they can be forwarded but gain credibility by how frequently they are forwarded.

But we're not here going to resolve the question of whether the Net is good or bad for knowledge. It is too intertwingly. Besides, we don't want to fall into the technodeterminism that believes that technology has only one outcome. We can learn how to use the Net to help us to know and understand our world better, or we can not. More important, we can teach our children or we can not.

Whether it is good or bad—whether we choose to respond in ways that make it good or bad—it seems clear to me that the networking of knowledge is working some fundamental changes in the nature of knowledge and long-form thinking's role in it.

First, although authority still is a stop sign, authorities are no longer primarily a special class of credentialed people producing a special class of works. Rather, authority is being defined more functionally: An authority is the last page in the linked chain you visit—the page whose links you choose not to click on.

Second, hyperlinked works establish an ecology of temptation, teasing us forward. When the temptations diverge from our aims, we think of those links as distractions. But we could just as well consider the new form of knowledge to consist of content that simultaneously settles an issue for us and baits our further interest.

Third, the authority of a work of knowledge is no longer a badge granted by its publication, but is continuously negotiated within the systems of editing, reading, reviewing, discussing, and revising that are now all aspects of one continuous and continual system.

Fourth, the edges of a topic are no longer marked by the gentle thud of the closing of a book. Rather, all topics, ideas, facts, and knowledge are embedded in webs of reference, discussion, and argument that put them to the test and to use.

Fifth, we have taken long-form works as the great achievement of human knowing because they have the luxury of developing ideas to completion. But now that ideas are freed of bound pages for their embodiment, it turns out that long-form works were never nearly long enough. They find order in the jumble of ideas they're clarifying, but they do so by imposing a discipline that keeps the reader's eyes focused directly on the path the author is treading. Released into the wildness of connected human difference, ideas foliate endlessly. There are no isolated ideas, and there never were; there are only webs of ideas.

Sixth, Sven Birkerts is right to point to the temporal nature of books. Paper books are published when they are done, and are done when they are published. They are the author's past tense: "This is what I wrote then." Web-form thought embeds knowledge in the present of conversation about that knowledge. The work undergoes revision even if its author never changes a word: If you are interested in Nicholas Carr's *The Shallows,* you will pursue that interest by exploring the fray of links it generates online; to refuse to do so would be to willfully ignore the opportunity to understand its meaning and its impact more fully.

Seventh, if we consider knowledge only as content apart from its new hyperlinked context of conversation, debate, elucidation, and denigration, we will miss the most important change that knowledge is undergoing. Knowledge now *is* the unshaped web of connections within which expressions of ideas live. It is no longer content that a lonely author conveys to readers sitting alone in their comfortable chairs, or that a professor standing at the head of the class conveys to students sitting in their uncomfortable chairs. If you want to know Darwin's theory of natural selection, you will find that this knowledge lives not just between the covers of his book, and not in any one head, and not in any one site. Knowledge now lives in the messy web that has grown around it, the way life lives not in our neurons, bones, blood and marrow but in their connection.

Finally, if books taught us that knowledge is a long walk from A to Z, the networking of knowledge may be teaching us that the world itself is more like a shapeless, intertwingled, unmasterable web than like a well-reasoned argument.

7

Too Much Science

In June 2010, National Public Radio's *Morning Edition* ran a story typical of the sort of science news that the media like to cover: A study found that if mice drink lots of coffee, they're less likely to suffer from little tiny cases of Alzheimer's.[1] Excellent! Have another cup because it's good for you!

Actually, you might want to put your mug down. While the story's pitch to listeners was that coffee prevents Alzheimer's, it—admirably—spent much of its time undoing that pitch. Coffee prevents Alzheimer's in mice! Maybe! Other studies on humans are provocative but inconclusive! There are other factors! We don't know! The mouse study that's the topic of this on-air story isn't really all that significant!

Then, perhaps because NPR recognized the problem, the very next piece was about how a different, tiny semi-scientific study got inflated into a cultural meme.[2] In the spring of 1993, a psychologist named Frances Rauscher played ten minutes of a Mozart piano sonata to thirty-six college students and then gave the students a test of their spatial reasoning. Rauscher also asked the students to take the same test after listening to ten minutes of silence, and again after listening to ten minutes of a person with a monotone speaking voice. The results of this experiment seemed clear, reported Rauscher: "[T]he students who had listened to the Mozart sonata scored significantly higher on the spatial temporal task."

The NPR story tracked how this modest experiment using a tiny, nonrandom group led to a small industry of Mozart for Babies CDs, Georgia's distribution of free Mozart recordings to every newborn in the state, and even death threats against Rauscher for reporting that she did not observe the same beneficial results from rock and roll.

The NPR piece attributed the inflation of this story to the American belief in self-improvement and the fact that we care desperately about our children. Of course. But it missed the primary cause in the sequence of events: Even before Rauscher found out her paper was going to be published, the Associated Press called her about it. Once the AP story went live, the Mozart Effect was everywhere. "I mean, we were on the nightly news with Tom Brokaw. We had people coming to our house for live television," Rauscher told NPR. "I had to hire someone to manage all the calls I had coming in." The headlines were along the lines of "Mozart makes you smart." Thus a handful of data gave rise to conclusions far beyond its reach.

Clearly, this is not supposed to be how science works, but it is how science embedded in the real world works. Some NPR listeners undoubtedly are to this day drinking an extra cup of coffee because a single experiment had an unexpected result. Thousands of babies grew up listening to cloying New Age renditions of Mozart's works because a statistically insignificant, nonrepresentative sample of college kids did marginally better at a narrowly defined task under poorly controlled circumstances. That's how science too often is taken up by our culture.

Of course, science *itself* doesn't work this way. The scientific method enables us to test hypotheses by isolating the causes of effects through carefully controlled, repeatable experiments. That's true for much of day-to-day science carried out in labs and workshops around the world. Even in scientific disciplines that are more theoretical or observational than experimental—evolutionary biology, for example—science has been a careful and conservative practice, patiently trying to tie facts together into theories that make sense of them.

That is an excellent strategy. But in some of the most important fields it's failing to scale. There is now more data than Darwin could have imagined. For example, both Thomas Jefferson and George Washington recorded daily weather observations, but they didn't record them hourly or by the minute. Not only did they have other things to do, such data didn't seem useful. Even after the invention of the telegraph enabled the centralization of weather data, the 150 volunteers who received weather instruments from the Smithsonian Institution in 1849 still reported only once a day.[3] Now there is a literally immeasurable, continuous stream of climate data from satellites circling the earth, buoys bobbing in the ocean, and wi-fi-enabled sensors in the rain forest.[4] We are measuring temperatures, rainfall, wind speeds, CO_2 levels, and pressure pulses of solar wind. All this data and much, much more became worth recording once we could record it, once we could process it with computers, and once we could connect the data streams and the data processors with a network.

And yet, as a culture, we can't handle an experiment done in a classroom on a spring day with thirty-six college students. How then will we ever make sense of scientific topics that are too big to know? The short answer: by transforming what it means to know something scientifically.

This would not be the first time. For example, when Sir Francis Bacon said that knowledge of the world should be grounded in carefully verified facts about the world, he wasn't just giving us a new method to achieve old-fashioned knowledge. He was redefining knowledge as theories that are grounded in facts. The Age of the Net is bringing about a redefinition of scientific knowledge at the same scale. What we've discussed so far in this book should lead us to hypothesize that scientific knowledge is taking on properties of its new medium, becoming, like the network in which it lives: (1) huge, (2) less hierarchical, (3) more continuously public, (4) less centrally filtered, (5) more open to differences, and (6) hyperlinked.

Let's look at each of these effects on science, to see how a discipline deeply devoted to truth is being affected by the networking of knowledge.

1. Too big for theories

In 1963, Bernard K. Forscher of the Mayo Clinic complained in a now-famous letter printed in the prestigious journal *Science* that scientists were generating too many facts. Titled "Chaos in the Brickyard," the letter warned that the new generation of scientists was too busy churning out "bricks"—facts—without regard to how they go together.[5] Brickmaking, Forscher feared, had become an end in itself. "And so it happened that the land became flooded with bricks. . . . It became difficult to find the proper bricks for a task because one had to hunt among so many. . . . It became difficult to complete a useful edifice because, as soon as the foundations were discernible, they were buried under an avalanche of random bricks."

If science looked like a chaotic brickyard in 1963, Dr. Forscher would have sat down and wailed if he were shown the Global Biodiversity Information Facility at GBIF.org. Over the past few years, GBIF has collected thousands of collections of fact-bricks about the distribution of life over our planet, from the bacteria collection of the Polish National Institute of Public Health to the Weddell Seal Census of the Vestfold Hills of Antarctica. GBIF.org is designed to be just the sort of brickyard Dr. Forscher deplored—information presented without hypothesis, theory, or edifice—except far larger because the good doctor could not have foreseen the networking of brickyards.

Indeed, networked fact-based brickyards are a growth industry. For example, at ProteomeCommons.org you'll find information about the proteins specific to various organisms. An independent project by a grad student, ProteomeCommons makes available almost 13 million data files, for a total of 12.6 terabytes of information. The data come from scientists from around the world, and are made available to everyone, for free. The Sloan Digital Sky Survey[6]—under the modest tag line "Mapping the Universe"—has been gathering and releasing maps of the skies gathered from twenty-five institutions around the world. Its initial survey, completed in 2008 after eight years of work, published information about 230 million celestial objects, including 930,000 galaxies; each galaxy contains millions of stars, so this brickyard may grow to a size where we have trouble naming the number. The best known of the new data brickyards, the Human Genome Project, in 2001 completed mapping the entire "genetic blueprint" of the human species; it has been surpassed in terms of quantity by the International Nucleotide Sequence Database Collaboration, which as of May 2009 had gathered 250 billion pieces of genetic data.[7]

There are three basic reasons scientific data has increased to the point that the brickyard metaphor now looks nineteenth century.

First, the *economics of deletion* have changed. We used to throw out most of the photos we took with our pathetic old film cameras because, even though they were far more expensive to create than today's digital images, photo albums were expensive, took up space, and required us to invest considerable time in deciding which photos would make the cut. Now, it's often less expensive to store them all on our hard drive (or at some Web site) than it is to weed through them.

That's why when Data.gov went up within a few months of the policy directive creating it, the keepers of the site did not commit themselves to carefully checking all the data before it went live. Nor did they require agencies to come up with well-formulated standards for expressing that data. Instead, it was all just shoveled into the site. Had the site keepers insisted on curating the data, deleting that which was unreliable or judged to be of little value, Data.gov would have become one of those projects that each administration kicks further down the road and never gets done.

Second, the *economics of sharing* have changed. The Library of Congress has tens of millions of items in storage because physics makes it hard to display and preserve, much less to share, physical objects.[8] The Internet makes it far easier to share what's in our digital basements. When the datasets are so large that they become unwieldy even for the Internet, innovators are spurred to invent new forms of sharing. For example, Tranche,[9] the system behind ProteomeCommons, created its own technical protocol for sharing terabytes of data over the Net, so that a single source isn't responsible for pumping out all the information; the process of sharing is itself shared across the network. And the new "Linked Data" format makes it easier than ever to package data into small chunks that can be found and reused. The ability to access and share over the Net further enhances the new economics of deletion; data that otherwise would not have been worth storing have new potential value because people can find and share them.

Third, computers have become exponentially smarter. John Wilbanks, vice president for Science at Creative Commons (formerly called "Science Commons," about which we'll learn more soon), notes that "[i]t used to take a year to map a gene. Now you can do thirty thousand on your desktop computer in a day. A $2,000 machine—a microarray—now lets you look at the human genome reacting over time."[10] Within days of the first human being diagnosed with the H1N1 "swine flu" virus, the H1 sequence of 1,699 bases had been analyzed and submitted to a global repository.[11] The processing power available even on desktops adds yet more potential value to the data being stored and shared.

The brickyard has grown to galactic size, but the news gets even worse for Dr. Forscher. It's not simply that there are too many brick-facts and not enough edifice-theories. Rather, the creation of data galaxies has led us to science that sometimes is too rich and complex for reduction into theories. As science has gotten too big to know, we've adopted different ideas about what it means to know at all.

For example, the biological system of an organism is complex beyond imagining. Even the simplest element of life, a cell, is itself a system. A new science called "systems biology" studies the ways in which external stimuli send "signals" across the cell membrane. Some stimuli provoke relatively simple responses, but others cause cascades of reactions. These signals cannot be understood in isolation from one another. The overall picture of interactions even of a single cell is more than a human being made out of those cells can understand. In 2002, when Hiroaki Kitano wrote a cover story on systems biology for *Science* magazine—a formal recognition of the growing importance of this young field—he said: "The major reason it is gaining renewed interest today is that progress in molecular biology . . . enables us to collect comprehensive datasets on system performance and gain information on the underlying molecules."[12] Of course, the only reason we're able to collect comprehensive datasets is that computers have gotten so big and powerful. Systems biology simply was not possible in the Age of Books.

The result of having access to all this data is a new science that is able to study not just "the characteristics of isolated parts of a cell or organism" (to quote Kitano) but properties that don't show up at the parts level. For example, one of the most remarkable characteristics of living organisms is that we're robust—our bodies bounce back time and time again, until, of course, they don't. Robustness is a property of a system, not of its individual elements, some of which may be nonrobust and, like ants protecting their queen, may "sacrifice themselves" so that the system overall can survive. In fact, life itself is a property of a system.

The problem—or at least the change—is that we humans cannot understand systems even as complex as that of a simple cell. It's not that we're awaiting some elegant theory that will snap all the details into place. The theory is well established already: Cellular systems consist of a set of detailed interactions that can be thought of as signals and responses. But those interactions surpass in quantity and complexity the human brain's ability to comprehend them. The science of such systems requires computers to store all the details and to see how they interact. Systems biologists build computer models that replicate in software what happens when the millions of pieces interact. It's a bit like predicting the weather, but with far more dependency on particular events and fewer general principles.

Models this complex—whether of cellular biology, the weather, the economy, even highway traffic—often fail us, because the world is more complex than our models can capture. But sometimes they can predict accurately how the system will behave. At their most complex these are sciences of emergence and complexity,[13] studying properties of systems that cannot be seen by looking only at the parts, and cannot be well predicted except by looking at what happens.

This marks quite a turn in science's path. For Sir Francis Bacon 400 years ago, for Darwin 150 years ago, for Bernard Forscher 50 years ago, the aim of science was to construct theories that are both supported by and explain the facts. Facts are about particular things, whereas knowledge (it was thought) should be of universals. Every advance of knowledge of universals brought us closer to fulfilling the destiny our Creator set for us.

This strategy also had a practical side, of course. There are many fewer universals than particulars, and you can often figure out the particulars if you know the universals: If you know the universal theorems that explain the orbits of planets, you can figure out where Mars will be in the sky on any particular day on Earth. Aiming at universals is a simplifying tactic within our broader traditional strategy for dealing with a world that is too big to know by reducing knowledge to what our brains and our technology enable us to deal with.

We therefore stared at tables of numbers until their simple patterns became obvious to us. Johannes Kepler examined the star charts carefully constructed by his boss, Tycho Brahe, until he realized in 1605 that if the planets orbit the Sun in ellipses rather than perfect circles, it all makes simple sense. Three hundred fifty years later, James Watson and Francis Crick stared at x-rays of DNA until they realized that if the molecule were a double helix, the data about the distances among its atoms made simple sense. With these discoveries, the data went from being confoundingly random to revealing an order that we understand: Oh, the orbits are elliptical! Oh, the molecule is a double helix!

With the new database-based science, there is often no moment when the complex becomes simple enough for us to understand it. The model does not reduce to an equation that lets us then throw away the model. You have to run the simulation to see what emerges. For example, Eric Bonabeau, an expert in models of this sort, suggests a simple game. Put ten to forty people in a room. Randomly assign each person two other people: an aggressor and that aggressor's prey. Each participant is given a single rule: Position yourself between your assigned prey and its aggressor. Would any predictable pattern of movement result? If so, what? The only way to tell is to try it, either with guests at your next party or via a computer simulation.[14] It turns out that—according to the simulation—a party-sized group of people will almost immediately form a tight cluster in the middle of the room.

This result is unpredictable before running the simulation, but it perhaps feels somewhat explicable after the model has revealed it to us.

As the rules of behavior become more complex, we lose that sense, which may be illusory in any case. For example, a computer model of the movement of people within a confined space who are fleeing from a threat—they are in a panic—shows that putting a column about one meter in front of an exit door, slightly to either side, actually increases the flow of people out the door.[15] Why? There may be a theory or it may simply be an emergent property. We can climb the ladder of complexity from party games to humans with the single intent of getting outside of a burning building, to phenomena with many more people with much more diverse and changing motivations, such as markets. We can model these and perhaps *know* how they work without *understanding* them. They are so complex that only our artificial brains can manage the amount of data and the number of interactions involved.

The same holds true for models of purely physical interactions, whether they're of cells, weather patterns, or dust motes. For example, Hod Lipson and Michael Schmidt at Cornell University designed the Eureqa computer program to find equations that make sense of large quantities of data that have stumped mere humans, including cellular signaling and the effect of cocaine on white blood cells. Eureqa looks for possible equations that explain the relation of some likely pieces of data, and then tweaks and tests those equations to see if the results more accurately fit the data. It keeps iterating until it has an equation that works.

Dr. Gurol Suel at the University of Texas Southwestern Medical Center used Eureqa to try to figure out what causes fluctuations among all of the thousands of different elements of a single bacterium. After chewing over the brickyard of data that Suel had given it, Eureqa came out with two equations that expressed constants within the cell. Suel had his answer. He just doesn't understand it and doesn't think any person could.[16] It's a bit as if Einstein dreamed "$e = mc^2$," and we confirmed that it worked, but no one could figure out what the "c" stands for.

No one says that having an answer that humans cannot understand is very satisfying. We want "Eureka!" and not just Eureqa. In some instances we'll undoubtedly come to understand the oracular equations our software produces. On the other hand, one of the scientists using Eureqa, biophysicist John Wikswo, told a reporter for *Wired*: "Biology is complicated beyond belief, too complicated for people to comprehend the solutions to its complexity. And the solution to this problem is the Eureqa project."[17] The world's complexity may simply outrun our brain's capacity to understand it.

Model-based knowing has many well-documented difficulties, especially when we are attempting to predict real-world events subject to the vagaries of history; a Cretaceous-era model of that era's ecology would not have included the arrival of a giant asteroid in its data, and no one expects a black swan.[18] Nevertheless, models can have the predictive power demanded of scientific hypotheses. We have a new form of knowing.

This new knowledge requires not just giant computers but a network to connect them, to feed them, and to make their work accessible. It exists at the network level, not in the heads of individual human beings.

But bigness is just the first property of networks the new scientific knowledge is absorbing.

2. Flatter

Charles Darwin was not exactly a professional scientist. He was not a member of a university or some other institution. He supported his scientific work through his travel writings, and later in his life through an inheritance from his father. He was, however, well integrated into the community of scientists, and was a member of scientific organizations such as the Royal Zoological Society, the Royal Society, and the Linnean Society.

Gregor Mendel was at a whole other level of amateurism. Unable to pass the qualifying exams to teach high school students, he worked for years in his monastery's garden, observing the peculiarities of generations of smooth and wrinkled peas. Today Mendel's name is almost always followed by the phrase "the father of genetics." During his lifetime, however, he was unrecognized and uninvolved in either the profession or the community of science.

Science has a long tradition of embracing amateurs. After all, truth is truth, no matter who utters it. On the other hand, if the manuscript Gregor Mendel sent Charles Darwin had been marked as coming from a prestigious university, Darwin might have cut the folded pages and read it.[19] If the self-taught math genius Srinivasa Ramanujan had not written to three Cambridge professors in 1912–1913, his lifetime of work—including the "Ramanujan Conjecture"—might have vanished without impact. If Carolyn Shoemaker, a "homemaker and mother," had not married an astrogeologist, she might not have become the best-known amateur comet hunter, and co-discoverer of the Levy-Shoemaker comet.[20] If amateur astronomers in the Philippines and Australia had not notified professionals about the two-second flash of light they independently observed when watching Jupiter, we might not have known that the flash was caused by a small comet or asteroid smashing into the giant planet.[21] Amateurs could succeed because there were institutional professionals who could validate them.

The Internet has not eliminated the need for credentialed, professional scientists. Nor has it completely erased the line between professionals and amateurs. It has, however, smudged the lines: more contributors, more confusing relationships, more and messier relationships that mix the social with the institutional. Where there once was a gap between the professional and the amateur scientist—a gap defined and maintained by the credentialing process—the Net is putting out tendrils to find every way across the divide.

The first Maker Faire was held at the San Mateo Fairgrounds near San Francisco in 2005. Twenty thousand people showed up to see "self-balancing two-wheeled vehicles, computer-controlled Etch-A-Sketches, biodiesel processing units, biologically-inspired multiprocessors, scratch-built RFID readers, wind-powered generators, networked citizen weather stations, ornithology research systems, flying pterosaur replicas, and hundreds of other projects," in the words of Mark Frauenfelder, the editor of *Make* magazine, which inspired the event.[22] In 2008, three times that number attended. Frauenfelder does not attribute this growth in interest to the Web directly. Rather, he says, in the past few years, "some of the folks who had been spending all their time creating the Web, and everything on it, looked up from their monitors and realized that the world itself was the ultimate hackable platform."

Maker Faire embodies the hacker ethic and aesthetics that have driven the Web and the culture of the Web. And there is no doubt that even for those who do not want to do the sort of science that involves both hacksaws and bags of marshmallows—marshmallow guns are a signature Maker artifact—the Web has been a godsend for the amateur scientist. So many sites, so many forums, so many YouTube videos.

But amateurs can do more than invent clever gadgets that make you laugh. Amateurs can now move science forward more easily than ever since science first became professionalized and institutionalized. This is, of course, easier in the areas of science that do not require building, say, a large hadron particle collider ($7.2 billion) or an international space station ($120 billion). We've already seen examples of amateurs like John Davis solving engineering problems—how to pump oil up from the *Exxon Valdez*—via contests. But amateurs are contributing in yet more structured ways. For example:[23]

- Volunteers at Galaxy Zoo, a science crowdsourcing Web site, have created what it claims is "the world's largest database of galaxy shapes."[24] Beginning in July 2007 it posted images of a million galaxies from the Sloan Digital Sky Survey and asked people to do a simple categorization of each one: spiral-shaped or elliptical? And if spiral-shaped, clockwise or counter-clockwise? In a year, it received 50 million classifications, including multiple classifications of the same galaxies, enabling Galaxy Zoo to error-check the reports. Having proved that the process works, Galaxy Zoo started a second project that asked more detailed questions. This information—publicly available, of course—has already changed some assumptions common among professionals; for example, it turns out that red galaxies are not always elliptical.

- eBird.org aggregates bird watchers' lists to build a database of bird migration. At CamClickr,[25] a project of the Cornell Lab of Ornithology, "citizen scientists" will be able to classify breeding behaviors to enable scientists to search and sort images of birds. And the NOAH site (networked organisms and habitats) lets anyone submit an iPhone photo of an organism, which it adds to its database of which species live where.[26]

- A month after a team of eighteen geologists and astrophysicists from Italy and Egypt announced in *Science*[27] that it had found an impact crater in the desert in southern Egypt, a lone Italian physicist found a crater in Sudan by using nothing more than Google Maps, an open access software tool she helped develop, and a popular open-source image editing program. Since there are only 175 craters on the Earth known to have been caused by objects from space, these are important discoveries. Wired.com pronounced, "The age of armchair crater hunting has arrived."[28]

Not all amateur science projects are crowdsourced. Rather than handing data out to humans to scan, Einstein@Home parcels data out to personal computers volunteered from all across the Net; the computers crunch the numbers when they're not being used by the owners. In August 2010, computers owned by three nonscientists discovered a new pulsar that—according to the article in *Science*—may be "spinning at a record-breaking speed."[29] This is such amateur science that the amateurs need not do anything more than donate unused computer time.

Other amateur projects address complex problems with odd skills peculiar to the human brain. Predicting how proteins fold (which accounts for many of their properties) is a notoriously difficult problem for computers because the number of possible combinations is astronomical. So, a combination of computer scientists and biologists created FoldIt, an online multiplayer game in which humans compete and collaborate to correctly fold protein structures. The only expertise required is the spatial sense baked into the human brain.[30] A study of

Foldit players found that humans outperformed the best of the computer algorithms in several types of problem.

Amateurs can crowdsource the processing of large volumes of data, they can donate computer time, they can use peculiar capabilities of the human brain, but they can also be involved in deeply personal ways. For example, Stephen Heywood was a twenty-nine-year-old 6'3" carpenter who was restoring his dream house when in 1998 he was diagnosed with ALS (amyotrophic lateral sclerosis), better known as Lou Gherig's disease. ALS typically leads to death in two to five years. His older brother, James, became frustrated with the pace of research, particularly the development of possible therapies that then go through extensive clinical trials. So, he left his job doing technology development at the Neurosciences Institute and moved from California to Massachusetts to create the ALS Therapy Development Foundation, which now has thirteen full-time scientists and $20 million in funding.[31] In 2004, James, his brother Benjamin, and a long-time friend—all graduates of MIT—founded PatientsLikeMe.com, a site that not only enables patients to share details about their treatments and responses but gathers that data, anonymizes it, and provides it to researchers, including to pharmaceutical companies. The patients are providing highly pertinent information based on an expertise in a disease in which they have special credentials they have earned against their will.

From bird watchers to ALS patients, it's hard to see anything except goodness in all these collaborative amateur efforts. Yet, they seem to make little difference to the structure of scientific knowledge and authority. The amateurs in these cases are performing simple tasks that require little scientific skill or training: Does this white blob in the sky look like a pinwheel or an oval? How did you react to the new medicine you've been prescribed? These amateurs are not tearing down the wall of credentials within which the scientific community lives. They are extending the apparatus of science. Are they nothing more than human sensors?

Where are the amateurs—the Darwins and Mendels—who are influencing science through the sort of contributions that credentialed scientists themselves make? The amateurs' articles tend not to make it into scientific journals, and they tend not to be invited to give talks at scientific conferences. Yes, there are some exceptions—Jeff Hawkins's theory of brain function,[32] Ray Kurzweil on the biology of aging, Stephen Wolfram's theory of everything—but they tend to be crossovers from technical or scientific fields in which they've been highly trained. It seems that amateurs had a bigger effect on the ideas of science in the nineteenth century than they're having in the Age of the Net. This is not totally surprising. As our scientific knowledge has grown and as our instruments have let us pursue ever more minute questions, the scope of individual scientists has narrowed. It is difficult and often impossible to become a specialist without being accepted into the scientific institutions that provide focused training and access to the necessary equipment. With that institutional access comes institutional credentials.

But the contribution of amateurs becomes more substantial if we look not only at what individuals are doing but at what networks of amateurs are contributing. For example, Arfon Smith, the technical lead of Galaxy Zoo, told me about the discovery of "green peas." It began as a joke in the discussion area of Galaxy Zoo about the green objects that showed up in some photos. After over a hundred posts on the topic, the amateurs at Galaxy Zoo realized that there was a type of astronomical object that the professionals had not noticed. "In mid-2008," said Smith, "they put together a portfolio and delivered it to us, and insisted that we pay attention." It turns out that "the green peas are important. We're just beginning to understand how."[33] That insight, and its development, occurred within a network of amateurs; had it been only a single person's observation, the importance of "green peas" would not have been noticed.

Indeed, the space within which credentialed science occurs is becoming dramatically and usefully more entwined with the rest of its environment. For example, on August 6, 2010, the mathematician Vinay Deolalikar sent to his colleagues a manuscript proposing a solution to a mathematical problem so notoriously difficult that there's a million-dollar prize for solving it.[34] The "P≠NP" problem had previously brought down some seriously brilliant mathematicians who thought they had it licked. This time, however, the person who originally formulated the problem emailed Deolalikar's solution to some of his colleagues, saying, "This appears to be a relatively serious claim to have solved P vs. NP." This highly authoritative endorsement made the paper go viral, at least within the limited world of mathematicians interested in such matters. Unfortunately, within two days, Deolalikar's paper suffered the fate of prior attempts. As the blog TechCrunch put it, "Both armchair and professional math pundits proceeded to tear it apart in comments sections and subsequent blog posts, finding major flaws."[35] The credibility of the person who first formulated the problem got Deolalikar's paper circulated, but the critiquing of it occurred without much regard for credentials.

Of course, amateurs could have participated in the evaluation of the proposed proof even before the Web. But, first Deolaliker's paper would have to have been approved by credentialed experts, and then it would've been published in a journal designed for professionals. The amateur critiques would have had nowhere to go, except back into the original journal as a letter to the editor—an uphill struggle for those without credentials. Now, the ecology is substantially different. All material is available to everyone, and everyone is able to feed back into the system instantly. Yes, that's an overstatement: Not all material is published online, not everyone has access to the Web, not everyone is comfortable posting replies, not all posted replies have equal visibility, and many are simply cracked in the pot. But note that in the foregoing list of qualifications, "Not everyone has credentials" did not show up.

The gated communities of science still exist. The exclusive journals are still exclusive, and academic science departments still take credentials very seriously. Most science still has to pass through narrow portals to get published. That is an inevitable outcome of scarcity: There are only so many pages in a printed journal, and only so many seats at the faculty high table. But the engagement of and with amateurs via the Net has become so widespread that we already take it for granted. It feels natural, as it should; it is natural for amateurs to be interested in science, and it is natural for scientists to want to engage with those who share their passion.

The result is not an entirely flat topology. Getting an article accepted by *Nature* still carries more weight than posting it on your personal blog, and having it referenced by one of the highly prestigious science blogs will raise its visibility and prestige. But so will getting it discussed in a web of blogs, passed around in emails, "liked" by Facebook users, or voted up at Reddit.com. An article's impact on scientific thought now is inextricable from the waves it makes in the social networks—formal and informal—of scientists, amateurs, and citizens.

Traditionally, it has not been that way. For example, scientific journals measure their significance through a scoring system first proposed by Eugene Garfield in 1955. Called the "impact factor," it has become the de facto measure of the authority of a journal—a score so important that "editors break open bottles of champagne if their impact factor rises by a tenth of a decimal point or burst into tears if it falls," according to Richard Smith, the former editor of the *British Medical Journal*.[36] The score is computed by counting the number of articles published in a particular journal during the previous two years and then dividing that into the number of times those articles were cited during the same period. It can be gamed, however. Smith writes: "Authors cite themselves and each other in 'citation cartels' in order to boost their impact factors, and some editors require authors to cite work in their journals in order to increase the impact factor of the journal."[37] But even if it worked perfectly, there is another problem: It works at the speed of print. "If I'm writing a paper today it takes a year to get it peer reviewed and another year to get it published," says Victor Henning. "The impact factor today reflects what was important two to three years ago."

Henning is one of the founders of Mendeley, a software application becoming increasingly popular among scientists that provides a different way of measuring impact. Henning and his co-founder were graduate students in London when they came up with the idea. "We realized that both of us had the same problem even though we were in different disciplines: managing an overload of information," Henning told me.[38] They each had hundreds of downloaded articles on their hard drives, usually in the popular PDF format. Why not write software that would not only automatically extract the bibliographic information but also allow users to annotate and highlight interesting passages? Why not also make the articles searchable? Why not automatically pull out key terms to make the searches more precise? And if you know what articles people are downloading and what sections they're highlighting, why not use that information to show trends in research and to guide people to articles they might not know about? They launched Mendeley in January 2008 and, twenty months later, had 450,000 registered users and 33 million articles.[39]

The effect of Mendeley is being felt outside of the population of Mendeley users because it can give a much faster view of what papers are mattering to scientists than can the impact factor. Mendeley can also break the influence down to a fine granularity. What matters right now to systems biologists? To climate scientists? To evolutionary biologists? As Henning says, "Are they scrolling through it? Highlighting specific sections? Giving it particular keywords?" The information Mendeley aggregates does not give special weight to those with weighty credentials. But Mendeley hopes to introduce another sort of lumpiness to its network: allowing the behavior and recommendations of those within your social network to count more heavily in what it suggests might be of interest to you.

Mendeley is a microcosm of the new ecology of networked science. It hasn't removed the importance of credentials. It hasn't elevated untutored amateurs to the height of authenticated experts. But it has provided an environment in which many more can participate more fluidly, new forms of credentials and authority are emerging, the connections among the social and the authoritative are becoming much more complex, and the map of authority is both far more granular and far messier. As we've seen is so often the case with the Net, when it cannot remove walls, it vines the wildness around them.

3. Continuously public

When your sixth-grade science fair project failed to show that sticking Flintstone vitamins into the soil helped potted plants grow, your teacher patted you on the back and told you that a negative result was just as important in science as a positive one.

Baloney.

Scientific journals rarely publish research with negative results. But, says Peter Binfield, editor of Public Library of Science [PLoS], "If you're a researcher starting on a topic, it's very useful to know what others tried and failed to make work."[40] He says that the classic examples are clinical trials of new drugs on human subjects: "Generally only the positive results get published," despite the personal cost to the human subjects. "The knowledge of what didn't work is incredibly useful, even though it won't advance the value of your drug company's shares."

Jean-Claude Bradley would agree. In 2005, he was an associate professor of chemistry at Drexel University[41] who was publishing research and being granted patents. But, he explained, "I wasn't having the impact I expected to have. . . . It was all very secretive, and most of the work my students and I did never made it into journals." That's because most of Bradley's work was trying out chemical compounds, and "Nothing happened" just doesn't make much of a scientific article.

So, he started a blog as a place to put the daily information usually recorded in paper notebooks by scientists. He had looked into using one of the existing electronic notebooks, but, he says, "[t]hey're designed to keep people out, whereas I wanted a system that would be hard to keep private." Eventually, he started a wiki and called what he was doing "open-notebook science."

His first open notebook, "UsefulChem," was designed primarily to record his lab's work trying to find chemical compounds useful in the fight against malaria.[42] "Most of the people who are sick with malaria don't have a lot of money, so the drug companies aren't attracted by the profit motive," Bradley explains. His lab started testing as many compounds as they could, recording the results in an open notebook that contained no heroic narrative, just daily results. He then started another open notebook that crowdsources the nearly endless question of which chemicals are soluble in which other chemicals. The result is a mammoth spreadsheet of interactions, most of which are nonevents. But because chemical reactions need to occur in a medium that does not react with them, information about those nonevents is crucial if you are trying to create new compounds cheaply.

Bradley points to the way his UsefulChem notebook intersected with an open notebook created by Mac Todd, a chemist at the University of Sydney working on diseases caused by parasites in the developing world. "We noticed that our reactions could be used to make the compound they're investigating," Bradley says. They noticed because that's what the record of "failed" experiments told them.

"Most of what's published comes after everything has been worked out and cleaned up," he says, "years after the notebook entries have been made." With open notebooks, science no longer is bound by traditional publishing's assumption that there is a publication date before which the work is private and after which it is public. Open and continuous science is becoming far more common, whether expressed through the relatively small number of scientists using Jean-Claude Bradley's open-notebook idea, the increasingly common open repositories of articles, or blogs that chart the daily work of scientists and laboratories.

The importance of this shift from the old private-then-public publishing model to a continuous *now* is visible in the confusion it causes in other spheres. Bradley points to a controversy over who discovered a dwarf planet called Haumea, as recounted in Alan Boyle's book *The Case for Pluto*.[43] It seems that the astronomer Michael Brown discovered a series of dwarf planets starting in December 2004 but kept them under wraps until July 20, 2005, when he posted a notice that he would announce the discoveries at a conference in September. On July 27, 2005, Jose Luis Ortiz Moreno filed a claim with the Minor Planet Center that his team had discovered one of those dwarf planets. Controversy ensued because Moreno had used some of Brown's published data—his telescope logs—to find the contested dwarf planet. Brown says that he'd assumed people wouldn't run with the data because generally it's not official until you've checked and analyzed it sufficiently to let you submit it to a peer-reviewed journal.

As Bradley writes in his post, this was a clash of conventions. The old model thinks of publishing as a point in time. Once a work has gone through that temporal portal and is published, credit and the accompanying authority are bestowed upon the author. But, in an open-science model in which the work is done in public and thus there is no one moment in which the work goes public, credit and authority become harder to bestow unambiguously. Bradley's post asks us to imagine that Brown's original data had been published as he gathered it, in an open notebook. On the very first day that Brown found significant data, everyone "would have had the opportunity to know that a very significant find had been made. There were still details to work out—and the Brown group might not be the first to do all the calculations to completely characterize the discovery." The continuous now of science means that it will sometimes be harder to know exactly who discovered what, because the discovery itself will result from a public collaboration that some of the collaborators may not even be aware of.

Individual scientists may not like losing this coin of authority, but science is undoubtedly better for it.

4. Open filters

"*Nature* and *Science* have a rejection rate of something like 98 percent," says Peter Binfield, managing editor of *PLoS One,* a free online journal begun in 2006 by its parent, *Public Library of Science.*[44] "A more standard journal rejects about 70 percent," and that leaves out the papers that scientists did not bother to submit because they assumed they would be rejected. At *PLoS One,* 90 percent of submissions receive initial acceptance. "We publish what is at the bottom of the heap, as well as papers that would have been submitted to *Nature.*" If a paper is rigorous enough to have been acceptable to a traditional scientific journal, it is accepted by *PLoS.* The great majority of *PLoS's* articles are sent out for peer review to two or more academics who feed comments back to the author. "We only use peer review to decide if a paper deserves to join the scientific literature. We don't use it to determine how important it is." Minor science, including negative results, is still science. At last your sixth-grade science teacher was right.

At *Nature,* says Binfield, "the entire system is run by in-house editors, who are ex-academics and look at every submission, and manage the peer-review process." That model simply does not scale: No commercial journal could afford to put most submissions through peer review. At *PLoS,* there are 930 academic editors. "If they see a paper that meets all the criteria, they can accept it without sending it out for external peer review." They rarely skip the peer-review process, although in some fields where time is of the essence, the process is more frequently short-circuited. For example, at *PLoS Currents Influenza,* a board of expert influenza researchers verifies that a submitted article is "a real science article from a scientist and that it's on topic." If so, the article is posted within minutes so that information can keep pace with the spread of the disease.

Binfield's motive in sending articles out for peer review is not so much to ensure quality as to reassure academics. "People wouldn't submit to a journal without peer review because their tenure committee wouldn't count it." He points to the "impact factor" that quantifies the influence of a journal. "It's widely recognized as a bad measure, but even though everyone knows that it's bad math and distorts science, people rely on it anyway." So, six of seven of the *PLoS's* journals restrict publication to papers that are relatively important. *PLoS One* does not. So long as it's good science, *PLoS One* will publish it.

Especially at *PLoS One,* this is changing the way people read. "People are going to have to have a different mindset," says Binfield. "There's a vast literature out there and people aren't going to spoon-feed you the best. You have to find it. That's a downside of our model." He pauses. "The upside is that everything gets published."

And not just published. Thanks to the open access movement, anyone with an Internet connection can read it. Open access advocates have been pushing to break the lock that traditional commercial publishers have had on scientific knowledge. The publishers have decided what gets published, they have limited access to what makes it through their filters by charging subscription fees that often range above $10,000 year, and they have drastically limited the circulation and reuse of scientific articles by keeping a tight hand on their copyrights. *Open access journals*—peer-reviewed, albeit occasionally in new, experimental ways—make their content freely available. *Open access repositories* provide a place where the public can freely find and read articles deposited by scientists and researchers at any stage in their work's development. Without these open access resources, the ecosystem would be deprived of air. Peter Suber, a center of the network of open access advocates, reports that "OA [open access] is growing fast through repositories and peer-reviewed journals. OA infrastructure is expanding, OA policies are spreading to more universities and funding agencies, and understanding about OA itself is improving among researchers and policy-makers."[45] For example, faculty at five of Harvard University's nine schools have voted overwhelmingly—unanimously, in two cases—to require faculty to post all articles accepted by closed access publications into an open access repository, although there is an easy exemption process. Open access publishing opens the spigot much wider, which means we need better ways to find what we need and to help us evaluate its worth. Already, sophisticated tools have emerged to enable social filtering and the computational discovery and ranking of content.

The authority that used to come from having been approved by a handful of anonymous peer reviewers at a major journal will increasingly be authority that comes from one's presence and place on the network. That is what peer review becomes when one's peers include all one's peers on the network.

5. Science with a difference

"I have a question for you," he said, taking out of his pocket a rumpled piece of paper on which he had scribbled a few key words. He took a breath: "Do you believe in reality?"

"But of course!" I laughed. "What a question!"

These are the first sentences in Bruno Latour's book *Pandora's Hope: Essays on the Reality of Science Studies.*[46] Latour is one of our deepest thinkers about science, a philosopher who bases his thought on in-place observations of scientists at work.

What a question indeed. But it's not hard to see how we got to the point where such a question can be raised. You can take the story of science over the past couple of generations as being about loosening the idea that science has a firm, unambiguous grip on the truth:

- Karl Popper in 1934 gave us a way to cleanly separate science from pseudo-science by telling us that a truly scientific statement is *falsifiable*—that is, there is a way to prove it false.[47] "Chewing gum quickly dissolves in human saliva" can be shown to be false, and thus is a scientific statement. "Chewing gum likes to be chewed" cannot be shown to be false and is therefore not scientific. What places a statement within the realm of science is not that we know it is true but that there is some conceivable way we could prove it to be false.

- Thomas Kuhn in 1962 published *The Structure of Scientific Revolutions,* which itself revolutionized our idea of science. Kuhn argued that science was not simply a progressive march of discoveries that build on prior hard-won discoveries, bringing us ever closer to the truth. Rather, it turns out that the questions science asks, the facts it takes as relevant, and the explanations it gives all occur within an overarching scientific "paradigm" such as Aristotelian, Newtonian, or Einsteinian physics. Simple truths appear—and are relevant—only within complex, historical systems of thought, institutions, and equipment.

- In 1968, James Watson published *The Double Helix,* an account of his and Francis Crick's discovery of the structure of DNA. This highly readable account scandalized many because it revealed that scientists are governed by personal ambition as well as by a desire to find the truth. Many accounts since then confirmed that there is a psychology and sociology of science. For example, in *And the Band Played On,* Randy Shilts recounts the shameful delays in the discovery of the retrovirus that causes AIDS due to rivalries among labs and government agencies.

- Since the 1960s, the group of thinkers loosely clumped under the label of postmodernism have challenged every construct of science. The great French thinker Michel Foucault tracked scientific "discourses"—very roughly like Kuhn's paradigms—through history to show that they form not on the basis of facts and evidence but because of the confluence of all that we term history, including power relationships.[48]

The overall lesson of the past half-century of thinking about science is clear: Science is not independent of human frailties and the contingencies of history. Disagreements among scientists may be due not to different experimental results but to changes in paradigms, or the clash of ambitions, or differences in discourses. Scientific disagreements can thus be much harder to settle than your sixth-grade teacher led you to believe. Now science is proceeding across a network itself characterized by difference and disagreement. So, how does science handle the disagreements that networking exposes?

The answers are different depending on whether one is looking at scientists arguing with scientists or at scientists arguing with the rest of culture. But both answers have something important in common: Networked science is learning to live with (and within) difference, rather than assuming that it can be driven out entirely.

When scientists disagree with scientists. Some types of differences among scientists are obviously helpful. For example, the Planetary Skin initiative—a collaboration between NASA and Cisco—aims at providing a platform for measuring, reporting, and verifying environmental information, working across the data "silos" maintained in the many disciplines and institutions that deal with climate issues.[49] Juan Carlos Castilla-Rubio, Planetary Skin's CEO, told me: "There are ten or twelve disciplines that have never had conversations with one another. Hydrologists, economists, engineers, water resource management."[50] They need, says Castilla-Rubio, "a common operating environment, which is network based, that lets them have conversations: 'Oh crap! I didn't realize that what happens to this watershed will affect the types

of crops the government is promoting that consume a lot of water.'"
He pointed to a project they did with the British Antarctic Survey that
touches on about fifty different disciplines. Cross-disciplinary differences can be complementary, of course, but Castilla-Rubio maintains
that substantive disagreements are important: "If there's no ongoing
disagreement, how can you generate trust?"

Interdisciplinary approaches have long been recognized as effective
in addressing complex problems. But as science becomes networked,
the complexity of just about everything becomes more obvious. Timo
Hannay of Macmillan Publishers and former publishing director of
Nature.com told me, "Every time we look a level down, it turns out to
be more complex than we thought it was."[51] The same is also true when
we look up a level, of course. The Net's ability to scale-up large and fast
makes it a more suitable medium for tackling problems as immense as
how to understand a cell's signaling system or how to predict climate
change. The Net allows plenty of room not only for all that data but for
robust disagreements.

But how about disagreements at the other end of the scientific
spectrum, where scientists are arguing about, say, the solubility of a
chemical: Either it's five grams per liter or it's some other number.
There's no possibility of shaking hands and saying "Well, I guess we'll
have to agree to disagree" about issues as specific as that. For those
sorts of disputes over particularities, science continues to use its well-
worn processes of experimental method, public debate, and authoritative institutions.

There is, however, a more difficult—and fruitful—middle ground
opened up by the networking of science. Aggregating databases is not
like pouring together two CostCo-sized boxes of Cheerios. As you
bring together data from multiple disciplines, or even from within the
same discipline, if the datasets don't think about their data in the same
way—even if it's something as simple as whether the data is recorded
in metric numbers or not—the computer has no way of knowing that
the pieces actually go together.

That's why the Science at Creative Commons group is working to
create linked systems that allow scientists to ask questions of multiple

databases as if they were one database. John Wilbanks, the head of the group, says that they currently have about forty different bundles of data sources, some with eighty to a hundred different systems for naming or classifying information within them. "There may be fifty or sixty names referring to any particular gene," he told me.[52] "A database of yeast genes has a lot in common with a database of fish genes, but they were written by different people who gave them different names. If you want to retrieve everything we know about a gene, you can't . . . unless someone tells the computer that all those different labels refer to the same thing." Says Wilbanks: "We've been going through these data sources and assigning unique names to entities" including genes, proteins, and gene sequences.

In the paper-based days, to make two sets of data consistent, the owners of the data would have to make yes-or-no decisions when there were disputes. So, taxonomists in the nineteenth century spent a lot of time arguing over how to classify worms, bats, and duck-billed platypuses. In the Age of the Net, the amount of information overwhelms the taxonomists, and we generally don't have institutions with the power to enforce such decisions. So, Science at Creative Commons adopts an increasingly common tactic: "Fork it," as Wilbanks says. "The gene named 'ABCD' is the same as the gene named '1234.' We publish the mappings so that people who don't agree with us can use what we've done. We don't have the bandwidth to disagree with people who disagree with us." So, Science at Creative Commons lets you use your own preferred names. People who disagree with the disagreers can always pop up a level and link the two different names so that their own computers can know that these two names point to the same gene, no matter what those other scientists think.

This way of maintaining and overcoming differences is enormously powerful. At its core is a concept of *namespaces*—a domain within which names are unique. The telephone system is a namespace consisting of unique identifiers that follow certain conventions about the number of digits, the use of area codes, and so on. License plates are created within state-by-state license number namespaces; the same number can be issued to multiple cars so long as it's clear which states' namespaces they occur in. Likewise, a single individual can have multiple identifiers—a Social Security number and a driver's license number—so long as each is unique within its own namespace. Two collections of genetic information can each have their own namespace, giving genes their own identifiers. These namespaces enable people to disagree about how to classify and name things, while allowing computer programs to pull information together from both of them, so long as the computer knows how to map names from one namespace to another. Namespaces allow the Net to knit together unimaginable quantities of data without forcing everyone first to agree on it all. Namespaces enable a fruitful difference.

This pragmatic approach hides a deep change. There was a time when arguments over the classification of plants and animals were arguments over the order of Nature, which was an expression of God's mind. The namespace approach acknowledges that it's more important that we be able to share data than that we agree on exactly how that data should be categorized, organized, and named. We have given up on the idea that there is a single, knowable organization of the universe, a Book of Nature that we'll ever be able to read together or that will settle bar fights like the *Guinness Book of Records.* No, you organize your data one way, I'll organize it another, namespaces and data model translators will let us benefit from each other's research, and we'll still be able to learn from one another's research.

This is pragmatism not only in the usual sense of the term but also in the philosophical sense, as espoused by William James, John Dewey, and Richard Rorty. Although it would be more accurate to talk about the philosophical *senses* of the term (like so many academics, philosophers make their careers by finding ways to disagree with everyone else in their discipline), pragmatism is generally marked as a rejection of the older idea that knowledge is a "mirror of nature," to use the title of a book by Rorty.[53] Instead, think of knowledge as a tool. It has value if it helps us accomplish an aim. As Rorty wrote, "Modern science does not enable us to cope because it corresponds [to reality], it just plain enables us to cope."[54] Different aims require different tools.

The pragmatism embodied in namespaces—and pursued by those who, like John Wilbanks, work to stitch together a usable commons of scientific data without insisting on universal agreement—lets science proceed faster than it ever did when we thought progress consisted in getting all scientists to agree on how the universe was organized.

When science disagrees with citizens. A Google search reports that *Nature* magazine has used the word "shitless" once in its over 140-year history. It makes its solitary appearance in an editorial in March 2010: "Ecologist Paul Ehrlich at Stanford University in California says that his climate colleagues are at a loss about how to counter the attacks [on climate change science]. 'Everyone is scared shitless, but they don't know what to do,' he says."[55] Advises *Nature:* "[S]cientists must acknowledge that they are in a street fight, and that their relationship with the media really matters."

"Does science's relation with the media matter?" sounds like the question "Do you believe in reality?" posed to Bruno Latour. The answer is so obvious that you want to know how someone could even come to such a question.

One way of answering is to look not so much at the change in science as a way of knowing as to the change in science as an *institution*—as a set of people, roles, policies, and behaviors that governs what goes on within its walls. Within those walls, scientists generally know how to evaluate claims and one another. They know that making unfalsifiable claims is not science. They know that a refusal to document one's laboratory methodology is not science. They know that certain foundational ideas have accumulated so much evidence that casually denying them puts one outside the grounds of the institution of science. When those institutional walls were well-defined—or, more exactly, *because* those walls were well-defined—scientists were a special breed. They had their work to do, and those outside the walls would listen respectfully to them because the walls so firmly divided scientists from nonscientists.

The institutions of science are not vanishing. Universities still confer degrees, funding agencies still provide grants, laboratories still accumulate equipment far beyond the means of curious amateurs. When someone pronounces about, say, physics, it still matters a great deal if that person is a senior scientist working on the Large Hadron Collider or is a self-taught hobbyist with a theory gleaned from some local blogs.

But you no longer need to be standing on top of a wall to make your proclamations. Indeed, because we for the first time have a single medium for information, communication, and sociality, science cannot stay behind its institutional walls.

The *Nature* editorial is evidence of the discomfort this is causing. The old response was embodied in Al Gore's strategy. *An Inconvenient Truth* is a masterful argument for Gore's point that not only is the Earth's climate warming, the change is mainly due to human behavior. But Gore has with a fair bit of success marginalized "climate change deniers." They are entitled to their beliefs but cannot claim that those beliefs are scientific, according to those who follow Gore's strategy. They are not even engaging in wrong science. They're outside the walls of science, though their howling can be heard within.

No one says this is an easy situation. There are many beliefs rampant in our culture that are about science but that are not scientific, and they need to be identified as such. Creationism (or, as it is now called, Intelligent Design) is not falsifiable, and is therefore not a scientific alternative to evolution by natural selection. But with the breaching of the walls of science, such beliefs cannot be silenced by proclamation.

That's too bad. There were important advantages when we granted wise elders the power to declare some beliefs true and some false—or at least to declare some to be scientific and some to arise from other forms of belief. Jenny McCarthy, a grade-B actress and former *Playboy* model, has sufficient media presence to convince a substantial portion of the population that vaccinating our children puts them at risk of autism. No children will avoid autism because of McCarthy, but some will die from avoidable diseases because of her ignorance.[56]

So, *Nature*'s editorial is right. Scientists need to enter the fray, because the mechanisms of belief have become detached from the means of knowledge. Science is not going to be able to reassert its old-style authority because it has lost the medium that enabled it to flourish: a one-way channel in which there were those who spoke and those who listened. Our new medium so unifies information, communication, and sociality that it's nearly impossible to keep the strands of that triple helix apart. Even after Al Gore's magisterial lecture has deservedly won an Academy Award and led him to the Nobel podium, it will be heard by people who share a world by talking with one another. The conversations about science are not kept apart from the ones about politics, entertainment, and our children. And as Kuhn, Watson, Foucault, Latour, and many others have taught us, science was never entirely apart from the culture, society, and politics within which it occurs. Yes, for much of the knowledge that matters to us, science provides a methodology that leads us to beliefs that conform our desires better to the world's implacable realities. Within the discourse of science, Jenny McCarthy is wrong and Al Gore is right—and science has the best claim to uncovering the truth about their issues. Nevertheless, even scientific knowledge exists in a messy web of humans where we make decisions—for better and often for worse—based not just on information and knowledge but within a social realm of social striving, personal interests, shared hopes, motivating emotions, and barely sensed stirrings. That was always the case, but the old medium, by giving more authority to credentialed institutions, promoted the illusion of near-uniform assent.

We are in this together. We need to continue to assert the truths that come from methods more likely to reveal them. But being right is not enough. Truth is not enough. It can't be because it is and always has been a product of a culture that is held together by more than knowledge. The new network makes that truth unavoidable.

6. Hyperlinked science

If electronic media were hazardous to intelligence, the quality of science would be plummeting. Yet discoveries are multiplying like fruit flies, and progress is dizzying. Other activities in the life of the mind, like philosophy, history and cultural criticism, are likewise flourishing.[57]

So wrote cognitive scientist Steven Pinker in an op-ed in the *New York Times* in the summer of 2010. It would be difficult to find scientists who would disagree with this assessment overall, although every scientist would likely point to some pain point: lack of funding, government ineptitude, media sensationalism. . . . Still and all, this is a great age for science.

But on paper (so to speak) it shouldn't be. Not only are we overwhelmed with data, the filters that kept bad ideas on the fringes are failing. Disagreements that used to be ruled out of court are now in evidence—disagreements among credentialed scientists as well as among the tutored and untutored laity. This should be a dismal age for science. How is science prospering in an era that unsettles better than it settles?

The basic answer is that the Net's rebooting of science has revealed that the old ways were more broken than we'd thought. In a phrase: Science had been a type of publishing and now it is becoming a network.

Charles Darwin gives us an unusually clear example of the inextricability of science and publishing. In 1836, Darwin returned from his voyages on the *Beagle,* having made the observations that would lead him to his theory of evolution. By 1838, he had a clear idea of his theory.[58] In 1842, he wrote a 35-page "sketch" in pencil, but kept it private. In 1844, he wrote a 189-page manuscript that he kept private, but that he instructed his wife to publish in case he died.[59] In the next fifteen years, he worked on barnacles, published eight books, fathered nine children, and corresponded frequently with colleagues. He also took up experimental science—testing, for example, how long snails could stay attached to a duck's foot, to see if that could explain their geographic distribution.[60] But he did not publish his theory of evolution.

Then Alfred Russel Wallace wrote him a letter.

Wallace was a young naturalist who, among other adventures, had watched from a lifeboat as the ship sank that contained all his work from four years in the Amazon. In 1857, Darwin received a letter from Wallace, followed by a correspondence that resulted in Wallace sending Darwin a 20-page manuscript that laid out essentially the same theory as Darwin's. Wallace had probably never met Darwin,[61] but he admired his work, and simply had to share his idea. Darwin was thrown into a tizzy. He wanted credit for having originated the theory, but he did not want to cheat Wallace out of recognition or, worse, let it look like he stole the idea from Wallace. So, upon the advice of two of Darwin's closest colleagues, a reading of Wallace's paper and two by Darwin was arranged for July 1858, in the Linnean Society headquarters in London. Neither scientist was in attendance: Wallace was in New Guinea, and Darwin's young son had just died.[62] The readings met with little reaction. Not until the Linnean Society published them in its journal did Darwin's paper begin to kick up interest among scientists. Encouraged, Darwin took his children to the Isle of Wight and began the book that, after thirteen months[63] of writing, would revolutionize science: *On the Origin of Species.*

This story of Darwin and Wallace is a famous narrative that has been told in various ways. Sometimes it is about how a gentlemanly society dealt fairly with the young pup Wallace by graciously according him near-equal credit. Sometimes it's about the slow course of deep thought, steeping across the decades. Sometimes it's about the effect of personality on science. But it is also useful to see it as a narrative about how the paper publishing model has silently shaped science.

From that perspective, this is a story about two realms, the public and the private. Scientists work in private, free from the need to commit to a hypothesis. The private realm is not solitary: Darwin was an avid correspondent. But the private space of science was a place into which one had to be invited. It is where the work of science is done.

Then, when the private findings were judged to be secure and the work was judged to be done, the scientist petitioned to have his work released into the public realm. The medium by which science was made public required stamping the ideas into paper, making them difficult to change. The paper medium also conferred authority upon that which became public, since the expense of paper and the limits on shelf space required publishers to act as expert filters. Without publication, science would lose its ability to build upon what it knows and thus would come to a stop. So, the scientific process could be said to include scientists, hypotheses, equipment, and publishers: Take away any one element, and science would not exist in recognizable form.

But publishing was not merely the last step in the work of science. The nature of publishing has marked the nature of science itself. Science aims at settling matters as far as possible—albeit with admirable humility—in part because it has relied on a medium that prints irrevocably on paper. Science as a career rests on the attribution of ideas to individuals in part because the publishing model enables decisive attribution. Because of the limitations of the medium, science is generally accomplished in article-sized chunks that are relatively self-standing.

"You need a narrative to write a scientific paper," says Jean-Claude Bradley, the founder of open-notebook science. "You can't just assemble a bunch of random findings and put them into a paper. Typically, a paper is a story." Within his discipline of chemistry, he says the basic narrative is: "You're trying to do something. You make a new reaction. This is what you can do with it." He's right; *Nature* won't publish a mere table of data unless it is of unusual importance. But, says Bradley, that story doesn't fit most of what scientists do. For one thing, "You get a lot of results that are ambiguous and you can't include them because they don't fit into the story." Bradley says most of the data in the "open notebooks" he's pioneered are like that. Even so, he says, the notebooks "could be useful to others, even just for learning what not to do."

For one thing, networked science can be more accurate because the system of trust-through-authority is changing. Bradley once assigned his students the task of finding five different sources for the properties of a chemical of their choosing.[64] The results were sobering. In one case a paper that had spent five months undergoing peer review at *Biotechnology and Bioprocess Engineering,* a respected journal, reported the solubility of caffeine in water as 21.7 grams per liter (gpl) and the solubility of the chemical extract of green tea (epigallocatechin gallate, or EGCG) as 521.7 gpl. By actually looking up the citation, Bradley's student discovered that it was at odds with the information in its cited source, which reported EGCG's solubility as 5 gpl. Probably through a transcription error, the number for caffeine (21.7) got appended to the number for EGCG (5), resulting in EGCG's solubility being reported as 521.7 gpl. But even that figure of 5 is suspect, for the chain of data in this case goes back through several more sources to a published experiment that, unfortunately, does not contain enough detail to allow for thorough judgment of its accuracy. Therefore, not only is the chain of published data demonstrably faulty, we don't actually know what the solubility of EGCG is. And that was only one example his students unearthed.

Bradley's point is not that all scientific data is wrong. Rather, it is that "[t]rust should have no part in science." We used to need trust because paper-based publishing breaks knowledge off from its source. Now, however, science—which has always had a network of inter-cited publications—occurs within a network of links. We create these links by hand, computers prowl the Web suggesting new links, and the surge of interest in the Linked Data format is making it easier than ever to create clouds of linked data just waiting for new uses. In this hyperlinked environment, we will continue to tell science's stories, but those stories will be embedded within a system of connections. We will click to see the data. We will click to have our computers compare disparate datasets, surfacing the anomalies and disagreements that will never be entirely driven out from the data of science or from its stories. We will click to read the commentaries from amateurs and experts. We will click to respond, react, correct, contextualize, question, support, or align.

This will have—is having—a profound effect on science and on knowledge itself. When science was a type of publishing, it aimed at producing knowledge that was—like a publication—broken off from its source because it was embodied in a physical thing with a life of its own. The new issue of *Nature* arrives on the desk of the scientist, and she sighs in relief. Her research is out there at last. If, heaven forbid, a truck were to hit her this morning, the knowledge wouldn't die with her. It now has a life of its own that can be tracked and weighed.

But now that science is becoming a network, knowledge is not something that gets pumped out of the system as its product. The hyperlinking of science not only links knowledge back to its sources. It also links knowledge into the human contexts and processes that produced it and that use it, debate it, and make sense of it. The final product of networked science is not knowledge embodied in self-standing publications. Indeed, the final product of science is now neither final nor a product. It is the network itself—the seamless connection of scientists, data, methodologies, hypotheses, theories, facts, speculations, instruments, readings, ambitions, controversies, schools of thought, textbooks, faculties, collaborations, and disagreements that used to struggle to print a relative handful of articles in a relative handful of journals.

So, Steven Pinker is right: Science is doing better than ever thanks to the Net. There is more information than ever. More of it is available than ever. Computers can discover patterns that humans would never have noticed. Commons are forming from clouds of Linked Data. Collaborative tools allow scientists to work together across all boundaries. Because of all this, we are able to investigate entire systems of nature—including simple cells—that were beyond us even a few years ago. The Internet has accelerated the pace of science. The Internet has broadened science and increased its reach. There are few scientists who would undo the Internet, or would move permanently to a remote, disconnected cabin in the mountains in order to avoid it.

At the same time, it seems incontestable that this is simultaneously a great time to be stupid. If you want to ignore the inconvenient truths of science, you can surround yourself with a web of ignoramuses who provide a sham system of misconstructions that make falsehoods seem as profound as truths. It's hard to tell exactly how much stupider we are as a culture thanks to the Net because the previous media tended to make hard-won truths globally available while keeping ignorance local: What got published generally was what made it through careful, albeit imperfect, filters, whereas the niggling falsehoods flourished outside the broadcast towers. Nevertheless, it seems undeniable that falsehoods now find a wider audience and lodge themselves in that audience more firmly than ever.

The traditional media—inevitably?—tend toward the sort of reportage that tells us this morning that caffeine prevents Alzheimer's and this afternoon tells us that, as the satirical newspaper *The Onion* reported, "Eggs Good for You This Week."[65] It's just too tempting for them to overhype a provisional finding, even when they may six paragraphs down note that the study was only on a few mice. Even if the headlines turn out to be correct, the media create the impression that science consists of a set of true beliefs. But networked science is essentially different. It is huge, perpetually at odds, pragmatic, coordinating differences via namespaces, always uncertain. That is, networked science looks much more like the scientist's view of science than the media's view.

We might reasonably despair that the media will always get science wrong because they will always be attracted to the dramatic headline. We should have hope, however, that the participants in networked science understand far better, by virtue of their participation, how science actually works. This reaches beyond those who directly participate in online science projects. The sense of scale and the inevitability of disagreement that the Net makes manifest to most of its participants are in fact characteristics of the eternally humble scientific outlook: The universe is vast and impossible to know perfectly.

At its best, however, our experience of the Net will educate more people about the real nature of science as a grand, continuing collaboration among fallible humans. At its worst, the Net will enable us to be more emphatically wrong about the methods, attitudes, and results of science.

Which will happen, the best or the worst?

Both.

8

Where the Rubber Hits the Node

Facing reality sounds simple—but it isn't.
—Jack Welch[1]

IN HIS MEMOIRS, JACK WELCH, the fabled CEO of General Electric and *Fortune* magazine's "Manager of the Century,"[2] tells of his decision to stop manufacturing nuclear power plants in 1981 because the Three Mile Island meltdown two years earlier had scared America out of the nuclear power market. Welch was new on the job. The dedicated, expert managers of the company's nuclear business objected, telling him, "Jack, you really don't understand this business." Welch writes: "That was probably true, but I had the benefit of a pair of fresh eyes."[3] GE exited the manufacturing side of the nuclear power business, while continuing to service its existing nuclear customers, very profitably.[4] Welch says in his autobiography that he told this story "over and over again in the first few years as a CEO" because it makes a crucial point: To render a good decision "[a]ll you had to do was face reality and perform."[5]

Of course, that makes facing reality sound simple, but, as Welch says in the next paragraph, "it isn't." If facing reality means knowing what's what, in the Age of the Net, there's more *what*—and more "what???"—than ever. Making a decision means finding your way through a dense thicket of claims, deciding what information to believe and which sources to trust.

So, when you have to flop one way or another—yes, there will be a domestic market for nuclear power plants or, no, there will not be—does the knife-edge of decisions make the tangled uncertainty of networked knowledge irrelevant? Worse, does the networking of knowledge make decision-making harder and riskier because it puts up so many contentious possibilities?

Since decisions are made by people with the authority to do so, we'll explore these questions in this brief chapter by looking at the changing nature of leadership, starting with a non-networked example that may help clarify what it means to say that, just as knowledge is becoming a property of the network, leadership is becoming a property less of the leader than of the group that is being led.

—

If you want to understand leadership, one good place to start might be a school that has the training of leaders at the core of its mission. So I visited Lieutenant Colonel Anthony Burgess at the United States Military Academy at West Point where he directs the Army's Center for the Advancement of Leader Development and Organizational Learning (CALDOL).

Burgess speaks three languages. The first is the clipped, acronym-filled patois of the military; the second, the disciplinary jargon of scholars of cognitive science and education that he learned while getting his doctorate of science in knowledge management. The third is the language of the Web.

As you listen to Burgess switch from one language to another, it becomes clear that the director of the Army's leading-edge leadership innovation center actually does not talk much about leaders, at least in any traditional sense. If you ask him directly about leadership, he responds by talking enthusiastically and learnedly about how to create effective units that can accomplish their objectives even in the unpredictable circumstances combat troops typically face.

It's not because Burgess is some New Age-y Internet guru. He's hard-core Army: a West Point graduate who completed Ranger School and led a unit of the 82nd Airborne Division for three years. He is wholly devoted to developing leadership skills that will enable Army units to accomplish their objectives. But the leadership required is not that provided by a single individual, no matter how steely his or her gaze is, but rather the ability of the team to stay determined and motivated, to find new ways to accomplish their goal as the situation changes, to switch to new goals if that's what's called for, to be resilient as members of the team are injured or taken up with other tasks. For Burgess, these are desirable characteristics of a team, not of any one individual. Leadership for Burgess is distributed throughout the team, so that leadership becomes a property of a unit the way robustness is a property of an organism.

Burgess does not explicitly think of it that way, and was surprised when I pointed out to him that he consistently answered questions about leadership in terms of the effectiveness of groups. But it is in fact a sign of his commitment to the soldiers he's training. Strong leaders are a means to an end. If that end—a team that accomplishes its objectives—is best achieved through distributed leadership, then that is what West Point will teach.

The change has occurred in part because the Net has made people more familiar with the benefits of connecting across hierarchical lines. For example, CALDOL grew out of an online discussion forum that Burgess and his colleagues created that does not display the rank of those posting to it. But much of the impetus behind seeing leadership as a property distributed across a team comes from the nature of the two wars the United States began in the first decade of the millennium. As Major Rob Stanton explained, "In today's world, it's not enough to be able to do the job of the person above you. You have to do 18,000 different jobs. You have to be able to manage water systems, run a town hall meeting, issue micro-grants, be politically savvy . . . and that's if you're a twenty-five-year old E5 [sergeant]."[6] The successful unit—the one best able to accomplish its objective—consists of soldiers who not only have a broad range of skills but know how to learn quickly and respond creatively. Each soldier takes the initiative, every soldier collaborates. While the soldiers will of course obey orders that come down from the hierarchy, the group as a whole has to have the characteristics that enable it to succeed in an environment that changes faster than the hierarchy can respond.[7]

Obviously the Army remains quite hierarchical above the level of the on-the-ground unit. But the separation of leadership from a leader, and the infusing of leadership into the group that is led, is likely to become even more widely adopted, for it has become a standard and effective way for networks to achieve seemingly unrealistic goals.

———

On April 16, 2007, at a little past seven in the morning, a senior at Virginia Polytechnic Institute and State University shot and killed Emily Hilscher, nineteen years old. When Ryan Clark, twenty-two years old, tried to help Hilscher, he was shot and killed, too. About two hours later, the murderer entered Professor G. V. Loganathan's classroom, killed him, and then shot eleven of the thirteen students there, killing nine of them. Ultimately, thirty-two people died, some while barricading doors so that others could escape.[8]

The initial Wikipedia page about the killings went up within minutes of the first reports and consisted of a single line stating that a fatality had been reported. It was revised seven times in the next fifteen minutes, and rapidly became a useful and reliable source of information that evaluated and summarized reports from the media at the scene.

These horrifying murders stirred up a national controversy over gun control laws. But the shootings spawned a controversy at Wikipedia of a different sort: Did each victim deserve a separate Wikipedia entry?[9] The question had nothing to do with judging the worth of the lives lost. Rather, the issue was whether those lives met the criteria for being included in an encyclopedia. Of course, the possibility of creating thirty-two separate entries would not have even entered the mind of a paper-based encyclopedia. But Wikipedia does not face constraints on its size. Instead the issue was whether Wikipedia best achieves its aim of being a great encyclopedia by being maximally inclusive or by enforcing rigorous standards for what merits inclusion.

The controversy over including the Virginia Tech victims occurred within a long-running debate about the biographies of living persons, or "BLPs" in Wikipedia jargon. These are some of its most ferociously disputed entries, since reputations and egos are at stake, and there have been awful cases of vandals strewing outrageously false information in some BLPs. Of course, the victims of the Virginia Tech murders were not living people, but many of the same questions were raised during a five-day debate on Wikipedia about whether the victims should even be listed and, if so, how much biographical information to include. Ultimately, the decision was made to list them with minimal detail: Ryan Clark is listed as "senior in Psych/Biology/English" and Emily Hilscher as "freshman in Animal Sciences." Only six of those listed have links to their own Wikipedia articles: five professors, and the murderer.[10]

The debate among "inclusionists" and "deletionists" at Wikipedia continues to this day. But the choice in this instance was made easier by a prior decision made by Jimmy Wales, the co-founder of Wikipedia and its titular head. In September 2006, Wales added some language to one of the wiki pages that serves as a policy manual for Wikipedia: "What Wikipedia is not."[11] To negative descriptions such as that Wikipedia is "not a publisher of original thought," is "not a soapbox or means of promotion," and is "not a crystal ball," Wales added that Wikipedia is not a newspaper "and especially not a tabloid newspaper."[12]

Wales spoke and the policy was created, just like Jack Welch deciding that GE will no longer make nuclear reactors. Except that it wasn't like Welch's decision all. Wales wrote in a public, editable place. His words were debated, and altered without his permission. And anyone could have introduced that policy, not just Wales. At the time I write this, Wales's actual words have been entirely replaced by a more specific, and less memorable, expression of what constitutes sufficient notability for inclusion: "[I]f reliable sources cover the person only in the context of a single event, and if that person otherwise remains, or is likely to remain, a low-profile individual, we should generally avoid having a biographical article on that individual.[13]

There's another way Wales's decision was unlike Welch's: He exercises his role as decision-maker reluctantly and as infrequently as possible. In fact, when I asked Wales for some examples of decisions he's made, none sprang to mind. Finally, after a pause, he talked about a controversy the community was in the process of debating, having to do with allowing people to edit locked pages (articles temporarily closed to changes because of persistent vandalism or overheated back-and-forth revisions) by introducing a review process. A poll of the community had been held, but now the poll itself was being debated. So, what was the decision Wales put forward as an example of his decisiveness? To hold a second poll to resolve the controversy about the first poll. And what was the iron in the fist that Wales brought crashing down? "For the most part, people thought, well, that sounds reasonable. Sometimes people say it doesn't sound reasonable, and they say, 'Well, it's Jimbo, so what are you going to do?'"[14]

At GE, Jack Welch's underlings might have said the same thing about one of his decisions: It's Jack, so what are you going to do? But they probably would have meant it differently. You can't do anything about Welch's decisions because the organization is structured to foreclose that precise possibility; the person at the top is the person in charge.

Over the years, Wikipedia has developed a set of policies and processes that enable the community—the network—to make and amend decisions. When the network cannot come to agreement, other processes kick in, including an arbitration committee, and then, rarely, the ultimate arbitration committee-of-one, Jimmy Wales. But those escalations up the chain of command are considered to be failures of the preferred system of bold action by individuals, reviewed and elaborated by the community. "In the early days, I made a lot of policy decisions," Wales told me, "but that's not really sustainable."[15] These days, most of his decisions are either essentially coin-tosses when the community is evenly split or judge-like applications of principles that all in the community accept. Wales's decisions come out of a community and after they're made they are re-interpreted and re-expressed by the community.

The distribution of leadership across a network is not happening just at Wikipedia.

In 1991, a twenty-one-year-old Finnish student, Linus Torvalds, posted a message to a Usenet discussion board noting that he was beginning development of a free operating system. By 2006, Linux was the second most widely used operating system (after Microsoft Windows), and Torvalds estimated that there were about 5,000 developers working on it worldwide.[16] The Linux community is far from flat, but its hierarchy aims at maximizing the autonomy of individual contributors while ensuring that high-quality, reliable software emerges. In a 1997 essay famous among developers, Eric Raymond compared the way corporations developed software to building cathedrals, while the Linux way was more like a "a great babbling bazaar of differing agendas and approaches."[17] But the Linux bazaar has a clear center; Torvalds, highly respected as an engineer, makes crucial decisions about the core ("kernel") of Linux. A complex network of contributors takes care of the rest of what traditional leaders do.[18] In one sense, this is nothing but traditional delegation of authority. Except that for participants it's far more like pitching in than being delegated to.

Indeed, the official Constitution of the community that produces a bundled version of Linux called Debian says explicitly: "A person who does not want to do a task which has been delegated or assigned to them does not need to do it."[19] Authority can still be wielded against rogue members of the community—say, someone who does not follow the agreed-upon quality assurance practices—ranging from public shaming to taking away access privileges, and, as with all large-scale collaborative projects on the Internet, the Debian community has procedures for adjudicating disputes. Nevertheless, these projects are not hierarchies so much as networks in which some nodes are more equal than others. Leadership is distributed across the network itself as far as it can be.

How far is that? Debian does indeed have a leader, chosen in yearly elections. Any developer can enter the running by posting her or his platform on the Web site, and by participating in a series of online debates and question-answering sessions. Voting for the leader and on other issues proceeds in a complex fashion designed to protect minorities and to keep the group from splitting over contentious issues. The leader's powers are limited, and she or he is supposed to make decisions about only those matters over which no one else has responsibility—very different from the hierarchical model. The circumscription of the leader's powers is part of a conscious effort to maintain maximum autonomy for the rest of the community. In his book *Cyberchiefs: Autonomy and Authority in Online Tribes,* Mathieu O'Neil writes: "Debian must reconcile the central notion of each developer's autonomy, and the respect for difference, with the constraints deriving from the production of a complex system with quality standards of the highest order."[20] So, the software is divided into modules "to give developers full administrative control over their packages or teams, in a mini-cathedral model." The developers have to follow strict guidelines for their modules so they can be easily integrated into the whole. Debian also requires members of the community to maintain identities, rather than contribute anonymously, because only those who have earned the trust of their fellows are allowed access to key resources that, in the wrong hands, could bring down the entire software project. Debian distributes leadership and decision-making as widely and smoothly as it can while still meeting its core mission of creating efficient, scalable, powerful, reliable, innovative software.

This distributing of decision-making can be taken further. Noel Dickover is senior new media adviser working with the State Department.[21] (He is also a world-renowned pumpkin carver.)[22] Dickover is personally dedicated to applying technology to global social problems, particularly in the developing world. A traditional approach would be to build a worldwide nonprofit organization that funds software developers to work on the problems. The World Bank and USAID do that, but, as Dickover puts it, the people who win the grants are the people who are skilled at winning grants—not necessarily a skill that comes with being an incredibly inventive and productive "hacker" (in the good sense of a software ninja). Instead, Dickover creates "micro-ecosystems" that connect organizers in local communities with software developers anywhere on the planet. "Instead of determining top-down what the problems are and how to solve them, and then giving money to some developers, this is a much more grassroots approach, with just a few outside participants" who are brokering the connections, Dickover explains. "You've empowered the local people. You've made them the lead. And you've made the network their mentoring community."

He gives an example of the sort of project that inspired him and the group he helped found, CrisisCommons.org. Efforts to relieve the suffering caused by the 2010 Haitian earthquake were hindered by the fact that the streets of Port-au-Prince had never been fully mapped. So, OpenStreetMap.org posted satellite images of the city on its wiki.

People from around the world, especially Haitians living elsewhere, started adding in street names. The map was, according to Dickover, "insanely detailed" within just a couple of weeks, and was routinely used by the World Bank, the United Nations, the US Southern Command, the US Marine Corps, the Coast Guard—"anyone who needed to get across town." By the third week, the World Bank was funding people from OpenStreetMap to train local Haitians in the use of GPS equipment to add more and more local knowledge.

Dickover wants to make this sort of partnership of local people with a distributed network of developers more routine, so that we don't have to wait for disasters to spur action. So, he has been organizing State Department–sponsored "TechCamps" around the world. For example, at the TechCamp in Santiago, Chile, at the end of 2010, people from Brazil and Argentina said they'd like to have a way to aggregate the data gathered by election monitors so they could get an overall picture of the situation on polling days. The requesters wrote up a "problem statement" and even produced a video explaining the issues. TechCamp passed these over to a loose coalition of developers called Random Hacks of Kindness, which sponsors weekend "hackathons" to write socially good software. Out of a hackathon held in Nairobi came an election monitoring system that was then deployed in Kenya, but that will also be used in South America and elsewhere. More TechCamps are planned for Indonesia, Moldova, Lithuania, and India.

The fact that Dickover's work seems to have little to do with leadership or decision-making is exactly why it's a useful example of both. Dickover and his colleagues are creating networks of local people with needs and social-minded hackers with the skills to address those needs. The decisions about who should take up which projects are made by the people with the most local knowledge—those on the ground who know the problems intimately and the developers who know the software possibilities intimately. It is a highly efficient and effective way to make decisions. It works because it is a network.

There are some elements of traditional leadership here, however. Dickover and his Crisis Commons co-founder, Heather Blanchard, spend much of their time organizing events and partnerships. Dickover is able to do this not primarily because of his position in the State Department. In fact, he was brought into the State Department because of the success of Crisis Commons. Dickover's power as a leader comes from his ability to form networks around his idea. This worldwide ecology of talented hackers and local social organizers is succeeding at solving some urgent issues by distributing leadership as close to the ground as it can. Dickover, Blanchard, and others lead only by enabling that ground to be a network.

———

This has everything to do with the change in our knowledge strategies.

A Jack Welch stands atop a pyramid from whence the big decisions flow. He consults his lieutenants, who in turn have consulted their subordinates. At each step up the hierarchy, there is a reduction of information: More details are stripped out, the strokes get broader and broader. If decisions are going to be made at the top, then filtering, reducing, and concentrating information is the only reasonable strategy. Even then, Welch's reliance on making decisions "from the gut" (the subtitle of his memoir) can be read as a recognition that GE is too big to know with the brain.

Meanwhile, with millions of articles in the English version of Wikipedia, nobody could expect Jimmy Wales, for all his vision and leadership qualities, to be an expert on everything. In fact, Wales is not even as much of an encyclopedia geek—that is, an expert on the nature and history of encyclopedias—as others in the community.

That's why when a problem is escalated to Wales, his decisions are often more like the Supreme Court's than a CEO's: He cites some accepted principle that governs Wikipedia. The network of Wikipedians is bound together by shared enthusiasm, an ethos, and constitutional principles. Wales does not wield power so much as work to keep the network on the same page.

———

Of course, the success of large, collaborative projects on the Net such as Wikipedia, Linux, Debian, and Crisis Common's networking of global software projects hardly means that corporations are going to switch over to networked decision-making. Governments require businesses to have a hierarchy of accountability, and there are times when you simply need someone to decide nukes or no nukes. And it's easier to move leadership to the Net when the manufacturing process, as it were, occurs entirely on the Net. Nevertheless, traditional corporations have something to learn from these network-based enterprises.

First, network decision-making scales up better than hierarchical decision-making does, at least in some circumstances. For example, in a 2006 paper, Peter Denning and Rick Hayes-Roth argue that responding to events the size of Hurricane Katrina requires "hyper-networks": multi-component organizations that are truly huge, spread out, and differentiated.[23] These hyper-networks do not have a hierarchical top, relying instead on distributed decision-making. Denning and Hayes-Roth could just as well be describing large Internet projects: If Wikipedia and Linux had to rely on centralized leadership, they would never have been built as rapidly or as well.

Second, network decision-making also excels when decisions require a great deal of local knowledge, which is particularly the case when situations are fluid and diverse, or the path forward is not yet fully known. It does not work nearly as well when the goals are set externally, there are many workflow dependencies, the processes are well understood, and contributors feel compelled to follow directives.

Third, network decision-making can motivate people where hierarchical, top-down decision-making would have the opposite effect. Hierarchies often scale by suppressing difference and disagreement. But as Denning and Hayes-Roth—senior academics at the Naval Postgraduate School with considerable practical experience—write: "The component organizations in hyper-networks . . . have adopted decentralized decision making, because it enables them to work for the common purpose without giving up their separate identities." The networked collaborative organizations we've looked at would be depopulated in an instant if the members had to give up their local identities and loyalties.

Fourth, when decisions are distributed throughout the network, more of the local knowledge can be applied. There is, of course, a balance here, for you do not want local leaders to make decisions that work against the good of the whole. This is one reason that collaborative networks often structure themselves into fairly autonomous modules (as with Linux and Debian): Local expertise can have more effect, with less risk to the whole.

Fifth, when decisions are made locally throughout the network, they are likely to express the interests of the local members who, typically, are volunteers. This is one way to make "corporate social responsibility" more than a bullet item on a corporate PowerPoint slide. Of course, all of the networked collaborative efforts we've looked at also have structures in place to ensure that local units or individuals don't stray too far off the agreed-upon course.

Sixth, hierarchical organizations that rest the pointy end of the pyramid on the back of a single human being are not as resilient as organizations that distribute leadership throughout a connected network. That is one main reason that Tony Burgess wants leadership to be distributed across an on-the-ground unit.

Seventh, hierarchical decision-making is completely of a piece with our traditional reductive strategies for dealing with a world that is too big to know. That's to be expected, since decisions are almost always really about the relative merit of competing streams of knowledge: Should you believe the local branch that's telling you it's too risky or the market analysis that says it's a goldmine? As the business environment becomes more complex, thanks to globalization and, yes, to the burgeoning Net, reductive strategies run an increasing risk of going wrong by missing the detailed contours of the local landscape. Network decision-making keeps decisions as local as possible—quite literally the case with the global organization Crisis Commons.

For these reasons, and also because the new generation is having its expectations set by its Net experiences, decisions within hierarchies will increasingly take on characteristics of decisions made by networks.

I began this chapter by asking if the network properties of knowledge— its containing differences and disagreements, its never being entirely done and settled—become irrelevant in the moment when a leader has to make a yes-no decision. And it is certainly the case that in most business environments, leaders at the top still have to make the critical decisions that take the company one way and not another.[24] Does traditional top-down decision-making remain unaffected by the changes that networked knowledge introduces?

A little bit yes, but mainly no.

Jimmy Wales certainly does have to make yes-or-no decisions at times. But those decisions cannot be understood if they are taken as a mere moment outside of their networked context. In a networked collaborative organization, a decision is part of a wave rippling throughout the structure. Each decision has a history, expresses multiple local interests, and is absorbed by the organization that makes the decision its own.

As old-style hierarchical decisions increasingly occur within enterprises enmeshed in the Net, those decisions are taking on some of the network's properties, even if the decision-makers don't explicitly recognize it. The CEO of General Electric could be entirely off the grid, but still GE's engineers, product managers, and marketing folks are out on the Net, exploring and trying out the ideas that affect their branch of the larger decision tree. After the decision, they will engage with the network to get feedback that may affect the execution of the decision. The organization's appropriation of the decision—the way it makes the decision its own—will be accomplished over its network, and will be visible on the network, including inevitably and to some degree on the vast public network.

The networking of knowledge therefore does not end at the lonely, non-networked desk of the Decider who has to flop yes or no. The moment of decision is now explicitly a node in a network from which it arises and through which it pulses. And that's for a clear reason: The network contains far more knowledge than any single leader could contain, tap, or manage. As organizations get bigger and become more essentially entwined in the Net, it will take a network to make the wisest decision.

9

Building the New Infrastructure of Knowledge

Over the course of the previous eight chapters, I am sure that there have been many issues and ideas about which we have disagreed. But one thing should be certain: We are in a crisis of knowledge.

If this were like a crisis brought about by a shortage of water, it'd be clear that the way forward is to find a method to once again have enough water. But in the crisis of knowledge we can't even all agree about what knowledge is, and thus what a solution would look like. So, it's understandable that we often try to get a foothold by starting with something simple: Are the changes in knowledge good or bad? Not to put too fine a point on it, we want to know if the networking of knowledge is making us smarter or stupider.

The question is difficult, and not just because the word "stupid" is being used as a blunt instrument. Even more difficult is the phrase "is making us," for it implies a technodeterminism, the belief that technology causes us to use and understand it in particular ways. In its most extreme form, technodeterminism says that the Net sweeps in and inevitably tyrants tremble, media cartels disintegrate, and collaborative castles rise in the air. Technodeterminism is no different for those who think that the Internet inevitably makes us stupid—rewiring our brains, as Nicholas Carr argues. Anti-technodeterminists such as the sociologist Eszter Hargittai and the social media researcher danah boyd point to the ways our social class, age, and subculture affect how we use the Internet and what it means to us. To some, the Net may be an electronic

Republic of Letters, but others feel excluded because they don't have the technical skills, the free time, or the aggressive personality so many Net forums favor. For entire countries, the Net is not an open marketplace of ideas so much as a source of carefully controlled propaganda.

On the other hand, there are some basic elements of the Net experience shared by almost anyone who encounters it through a Web browser. These shared experiences are not inevitable, nor does everyone react to them the same way. Yet, they seem to have some likely effects that bear directly on how we understand knowledge. If you are a Westerner who has used a browser to experience the Net, you've probably come away with at least these ideas:

- *Abundance.* There is more available to us than we ever imagined back in the days of television and physical libraries.

- *Links.* Ideas can be hyperlinked, and you can go from one to another with a mere click.

- *Permission-free.* The default is that people can read, post, and build what they want on the Net.

- *Public.* What you can see, generally others can see. The Net is a vast public space within which the exclusion of visitors or content is the exception.

- *Unresolved.* The longer you spend on the Net, the more evidence you have that we are never all going to agree on anything.

Let's look at each of these basic, generally shared lessons from the Net, asking whether they make us better or worse knowers—and how they affect the nature of knowledge itself.

Abundance

If we define abundant as "more than we could ever use," then the knowledge in books was abundant before the Internet. Even before books, the hundreds of thousands of scrolls in the Library of Alexandria were more than could be carried out to safety from the great fire, much less be read in a lifetime. Only about 2 percent of the Harvard University library system's physical holdings circulate every year, and most of those are the same works that circulated the previous year.[1]

The new abundance makes the old abundance look like scarcity. The Google book-scanning project alone has over 15 million scanned books, which you can search through more easily than you can look up an item in the index of the book on your night table.[2] Harvard's Robert Darnton, whom we met in Chapter 6, is among those proposing a Digital Public Library of America,[3] a call that has excited interest among public and research librarians, the government, and some large Internet projects. Whether or not that project takes off, the Internet is likely ultimately to contain most of what's currently in libraries, with the exception of some private or difficult to digitize collections. And that's just for starters. Add to the content of our combined libraries the trillion pages or so already online, and now you're talking about abundance.

An infrastructure of knowledge that gives us access to more of the world's works than ever before—with the existing imperfect but amazing set of search tools—certainly seems better for knowledge-seekers than one that gives access to fewer. That is not the end of the story, for we also have access to more untruths, but it is a pretty promising beginning to the story.

If you are a researcher, your work has already been changed by the presence of this superabundant online world. If you cannot find the article you want because the journal has not digitized it, or because the journal wants to charge you $35 for access to that single item, you'll try to find a different article, if you possibly can. Call the decision not to track down the hardcopy in a library laziness if you want—and there are times when it will have bad consequences—but it feels like efficiency. Further, the ease with which we can explore means that with a click you can browse back from the sixth edition of Malthus's major work to discover that the first edition was almost entirely devoid of statistics—a task that probably would have been impossible in your local library. Our new apparatus vastly improves the ability of researchers to wander gently or drill in hard. Whether that new facility merely makes researchers facile is up to the researcher.

Our new infrastructure doesn't just open up an abundance, however. It makes the abundance apparent to us, and this changes how we understand knowledge. Although we can see only one screenful of the abundance at a time, that screen is full of links out, and we know that the links will lead to more links. Many of the screens trumpet the plenitude behind them: The results page for a Google search on "abundance" leads off with "About 40,500,000 results (0.27 seconds)." The inclusion of the retrieval time not only boasts about the search engine's prowess, it tells its users that the abundance is available, at hand. What could you accomplish in a traditional library in 0.27 seconds?

Of course, most of what the Net makes available to us does not count as knowledge. Knowledge at its most stately in our culture—the knowledge that advances us, that makes us most proud, that represents the best we can do as a species, that we want to spell with a capital K— has been the opposite of abundant. It has been rare and hard-won, such as Darwin with his barnacles or Jonas Salk with his vaccine. This vision of Knowledge as a carefully curated realm is contradicted by the wild web every hyperlink invites us into. We seem to be making the cultural choice—with our new infrastructure's thumb heavily on the scale—to prefer to start with abundance rather than curation. Include it all. Filter it afterward. Even then, the filters do not remove anything; they filter forward, not out.

In this world of abundance, knowledge is not a library but a playlist tuned to our present interests. It is not eternally truthful content but subject matter good enough for our current task. It is not a realm but a path that gets us where we're going.

Links

Knowledge has always occurred within a context developed through some form of network and maintained through some form of links. *On the Origin of Species* may not be footnoted, but it points to objections raised by others, and was written by a man embedded in a social network of colleagues and opponents. Modern printed works of scholarship footnote everything they can, in part to authenticate ideas but increasingly to avoid transgressions in an insane economy of micro-ownership of ideas.

When knowledge was communicated and preserved on paper, it had to work around the fact that connected ideas were expressed in a disconnected medium. You knew that few people would track down the works referenced in your footnotes, so you had to pull into your text as much of the referenced work as you needed (begging the permission of the jealous god Copyright). You therefore became the spokesperson for others in your network of knowledge. You would do your best to be fair, but you knew that you were reducing your fellow scholar to the excerpt you chose. You had no alternative. It's not as if you could fit that other book inside your book.

Now you can. You will still include the relevant passage from the referenced—linked—work, but you will do so aware that your reader can instantly check it and read more than you've included. Links erode authorial control.

Links also change the basic topology of knowledge. People will continue to write long works because complex knowledge needs time to develop, just as any narrative does. But readers are being trained by the linked Net to see any piece that develops an idea as living within a connected, traversable web. We know that every topic stretches beyond its covers because we see links penetrating pages a thousand times a day.

Links are subverting not just knowledge as a system of stopping points but also the credentialing mechanism that supported that system. Credentials will continue to count, especially when the topic is important and capable of being reliably settled—for example, advice about diabetic diets versus opinions about comfortable shoes. In the appropriate cases, we can only hope that our children and fellow citizens will pay attention to credentials. But even where credentials count for much, the credentialed works exist in a network where recommendations and the opinions of others also count for something. For one thing, you probably got to the credentialed site through recommendations from uncredentialed pages.

We created knowledge as a system of stopping points both because that's what paper enabled and because it's a highly efficient strategy. Our sources in the paper-based ecology may turn out to be less trustworthy than we'd hoped—anyone know the real solubility of EGCG, hmm?—but with no stopping points we wouldn't get anywhere. The linked infrastructure we've built for ourselves also provides stopping points, but with an implicit statement that there's more there. The last word is now never the last word.

Thus, the links that we all encounter in every encounter with the Web thoroughly transform the shape of knowledge, the role of authorities and credentials, and the reasons and places we allow our inquiries to stop.

Permission-Free

The first two characteristics of knowledge's new infrastructure seem well-aligned with what we've taken to be "what knowledge wants," to modify the title of an excellent book by Kevin Kelly.[4] Who could complain about there being an overabundance of knowledge that is easily traversable via links?

The Net being permission-free, on the other hand, feels like a challenge to traditional knowledge. Knowledge has been like a club that accepts new members—a book, an article, an idea—only after they've been examined by a credentialed board of experts. Let anyone publish whatever they want, let anyone curate a collection just by putting together a few links, and the Knowledge Club loses value. While the Net is not completely permission-free—you are constrained by the laws of your land, plus there is the "silent permission" that leisure time and money provide—knowledge has lost its exclusivity.

Within the permission-free ecology of the Net, knowledge clubs—curated sites that give us reliable information—continue to exist, for which we should be thankful. If anything, we are seeing many more such clubs on the Web. But this proliferation of clubs with different criteria for admission is beginning to play havoc with our institutional ways of deciding who is an expert. In a networked knowledge ecology, how does a tenure committee decide how to weigh four peer-reviewed books against

12,045 tweets and 3,754 blog posts? How about the scholarly arguments applicants are engaged in within the comment sections of other blogs?

The problem is that when there were few clubs, we knew what they stood for: Getting an article published in *Nature* would definitely go to the top of your résumé. In an abundant, permission-free publishing environment, metadata—information about information—becomes more important than ever. The difference between the sentence "Birds descended from dinosaurs" when it comes from some anonymous stranger on the Internet and when it comes from *Nature* is the metadata that says *Nature* is reliable. It used to be that the authority metadata was implicit in how the knowledge was distributed: It came from *Nature,* or from your physician's mouth. The mere existence of a book from a reputable publisher was metadata that established that at least some authorities thought it was worthwhile. Since simply being posted in a permission-free world conveys zero metadata about authority, that metadata now has to be made far more explicit. So, many sites have sophisticated metadata trust systems, tuned to particular sites and needs—Amazon reviews, your bank page displaying the last digits of your account number to vouch for the page's authenticity. The web of links itself can be evidence of authority, although of course we can also be misled, as ever was the case.

Architecturally, the Net's lack of an overall permission system makes knowledge less like self-standing content—bricks, in Bernard Forscher's sense—and more like nodes that cannot be fully credited or even fully understood outside of the network that connects them.

Public

There was a time when we thought we were doing the common folk a favor by keeping the important knowledge out of their reach. That's why the Pope called John Wycliffe a heretic in the fourteenth century for creating the first English-language translation of the Christian Bible.[5] For a long time, we taught children only enough to get them out into the fields or factories. Now we are providing the public not just with an education and a local library but with one-click access to a near-infinity of works of knowledge and culture.

In fact, we've not only given everyone access to these works, we've created a new type of public space for them. The pages within this space accrete links, each leading to some reflection on the work's meaning and value. The web made by those links has its own meaning and value, which can be mined and fed back into the network. Putting a work into this new public space contextualizes it in a way that shelving a book in a physical library does not.

The cost of publishing to this new, permission-free public space is so low that frequently we post not only the finished work but the drafts and perhaps even the editorial discussions that shaped them. We can all see the Sausage of Knowledge being made, one link at a time. We can see that what used to come forward as the canon of knowledge depends upon the filters we're choosing and how we apply them.

We used to think that knowledge is what is true independent of us. Now we are faced with the fact that knowledge is not a mirror held up to nature but, rather, a web of connections that shows itself to us depending on our starting point, viewpoint, and inescapably human sense of what matters to us. We had hoped that knowledge is independent of us. Now we know for sure it is not.

Unresolved

Geometric proofs have been exemplars of knowledge from knowledge's beginnings because, given their premises, we can be certain of their conclusions. Over the course of Western philosophy, we kept ratcheting up the standards of certainty, until we set Descartes to sitting alone in a room (wanting to avoid the distractions of everyday life is not new to the Internet era), making us wonder if even the knowledge that we exist might be the result of a malevolent god tricking us. If we can't be certain of something beyond the whiff of a glimpse of a tremor of a shadow of a doubt, then we do not know it. Or so says Descartes and the tradition he influenced.

The continuous ratcheting up of the standards of certainty seemed as inevitable as the increase in intrusiveness of airport security devices. But then nineteenth-century philosophers suggested that perhaps some knowledge was so ungrounded in reason that it could be held certain only with much fear and trembling (Kierkegaard). Perhaps knowledge's certainty panders to the weak-souled, and hides greater, more terrifying and joyful truths (Nietzsche). Perhaps the carefully constructed rational knowledge we've taken as the way to truth is based on a lived experience of a particular time and place of a creature that knows first of all that it will die (Heidegger). Perhaps what is known via science depends upon paradigms that determine what questions to ask and which answers count (Kuhn). Perhaps the idea of knowledge itself is an instrument of power wielded to retain the privileged positions of those who get to decide who and what makes it into the house of knowledge (Foucault).

We have lived through enough fundamental revolutions of thought to suspect that we don't happen to be living in the age that finally gets everything right. Yet, at one end of the knowledge pool, our general belief remains that to know something is to have driven out doubt. The sentence "I know where I live but I'm not sure about it" doesn't make sense. Parallel to this, we retain a fundamental sense that knowledge consists of the truths about which all reasonable people ought to agree.

Our new medium of knowledge, however, can't keep information, communication, and sociality apart. Post something and you may see colleagues and fellow travelers who almost agree with you, ridiculous conclusions drawn from it, and denials of even your most obvious points. You will experience what you have always known: There is nothing you can say to convince some people. The old Enlightenment ideal was far more plausible when what we saw of the nattering world came through filters that hid the vast, disagreeable bulk of disagreement.

The conspicuous inevitability of disagreement does not change everything. A scientist researching the solubility of chemicals will continue to use the same equipment and techniques. The historian investigating the role of the sugar industry in the slave trade is going to make the same travels, page through the same ledgers, and document her work in much the same way as before, albeit taking advantage of online sources. Nevertheless, the publicness of the Net has now made a pragmatic truth unavoidable:

What we have in common is not knowledge about which we agree but a shared world about which we will always disagree.

The Strategy of Too Much

No one knows how this will turn out. For one thing, we are not done innovating. For another, we don't know if the Internet is going to remain an open field for research, speech, and creativity.

The hopeful scenario is that we will repeat the pattern history has shown us. Literacy, printing, paperback books, and television all resulted in the vulgarization of prior forms. The uneducated and barely interested wasted their time and coarsened themselves with these tools. Yet we advanced. In part, we discovered value and treasures among that which we wrote off as vulgar. But we also advanced because the disciplines dedicated to discovering knowledge and plumbing its depths have moved the entire culture forward. The leading edge drags the trailing edge slowly behind it—what the Leonardos learn and create eventually improves the lives of everyone. This is not inevitable, but it happens. We can hope that it will happen more swiftly and evenly now that knowledge's medium is more accessible than ever before.

But hope is not enough. Suppose we were to take the question "Is the Net making us stupid?" to have less in common with questions such as "Is it raining outside?" and more with "Is my preferred political party going to lose the next election?" That second question is best answered not merely with a prediction but with a conditional: "Yes, unless you get up off your duff and do something about it."

What can we do? I have been maintaining throughout this book that knowledge is becoming a property of the network, rather than of individuals who know things, of objects that contain knowledge, and of the traditional institutions that facilitate knowledge. So, let's look at what we might do to help this network of hyperlinked superabundance be a better environment for knowledge.

What I've argued about technodeterminism suggests a general approach. If within our culture the Net generally shows itself one way more than another, if there are lessons we can expect virtually everyone to learn from participating on the Net, then the best way to move forward is to embrace the peculiar properties of this peculiar network. For example, don't try to reduce the network's inherent abundance by introducing artificial scarcities, such as imposing on digital libraries all the limitations on access inherent in physical libraries. Or, if the network within our culture has an inherent preference for weak ties (as opposed to deep, strong, abiding relationships), embrace the fact that weak ties can extend the reach of knowledge.[6] This type of jujitsu—moving into the punch— has the advantage of putting to good use the Net's strongest tendencies.

So, if we want to take advantage of the new strategy of knowledge that includes rather than excludes, and if we want to make use of the Net's own tendencies whenever possible, how might we address the challenge to act that we should hear in the question "Is the Net making us stupid?"

Here are five ways we can help make the networking of knowledge the blessing it should be.

1. Open up access

It's one thing to say that the exigencies of paper require publishers to carefully select what they commit to print. It is another for publishers to price their works so high that only an elite can read them. When academic journals are charging $20,000 for a subscription, they have become obstacles to knowledge rather than enablers. This is too high a price—and not just monetarily—to pay, which is why the success of the open access movement is as close to inevitable as economic and social trends can be.

Open access *journals* are how our knowledge system would work if we hadn't started it out on paper. Their articles are peer-reviewed, more people can learn from them, and the information they contain is available faster. Open access *repositories* provide a public place where scholars and researchers can make their work available at any point in its development process. These repositories make information and ideas available even sooner, and give access to the occasional worthy thought that did not make it through peer review. We should support them.

An open ecology is important also because it facilitates a policy of including everything and filtering afterward. If we filter before publication, we have to rely upon the best estimates by others about what will matter to readers. No matter how expert the custodians are, they cannot possibly anticipate every human interest or every turn of history. For example, before 2008, no responsible curator could have known that mentions of library issues in the minutes of Wasilla City Council meetings in 1996 might be of national interest and importance.[7] By giving users tools that let them filter after publication—anything from a search engine to sophisticated personalized navigation systems—we enable them to find information in the ways that make sense to them, and to access that which might otherwise have been deemed not worth preserving.

In an open ecology, we still benefit from the decisions made by library committees, journal editors, book publishers, and other custodians. We benefit more from the intersections of all these filters. We are learning how to mash up filters, and to filter the filters, and to use some filters as counter-indicator filters. For example, the Harvard Library Innovation Lab that I co-direct has built a program for browsing the 12 million works in Harvard's library system—a program that uses circulation records, course reading assignments, student ratings, and how often a book is called in early from a loan as indicators of the work's relevance to a community of scholars and learners. But because no single ranking can suit all needs, the program will let users assign their own weightings to these factors. As I write this, we are in the process of pulling in similar information from other libraries so that the filters can be more finely adjusted. Filters are becoming as explicit and public as the objects of knowledge have been. Getting good at filtering makes the open Net more useful to you, and increases its value.

There are also political ways in which we can support openness. Around the world the Internet is under attack by governments and powerful commercial interests. For a mix of practical and venal reasons, these entities—especially Internet access providers and their friends in government—are making a concerted effort to reduce the Net, preferring some sorts of content over others, limiting access, stripping out anonymity, and treating the Internet as if it were primarily a way of delivering sanitized, safe commercial content. We need to be vigilant.

Then there are the restrictions imposed by copyright that are ratcheting up even as the opportunity to benefit from open access has increased. Since 1989 in the United States all works are automatically copyrighted, even if you don't want them to be. That's why Lawrence Lessig (who you may remember from Chapter 2 worrying about some unwanted effects of open government data) and a few others started Creative Commons in 2002. CC, as it is known, makes it dead easy to declare that you are fine with others using your work without first asking permission. Hundreds of millions of items now bear the CC mark. Until we can get changes in the copyright law—sometime before the sun burns out, we hope—CC'ing your work and supporting CreativeCommons.org will help develop an open ecology for knowledge. The new age of open access should have us learning from the wisdom of the founders of the United States, who saw copyright as a necessary temporary restriction on access that should last a reasonable period (fourteen years back then, seventy years after the death of the rights holder now) before a work would enter the public domain—a sensible balancing of incentives for creators and the good of an educated, creative public.

2. Provide the hooks for intelligence

The strategy of abundance has two main risks: First, we won't find what we're looking for. Second, we will find lots of appealing stuff that panders to our lowest desires. A single practice addresses both concerns, although imperfectly.

The solution to the information overload problem is to create more information: metadata. When you put a label on a folder, you're using metadata so that you can find the papers within it. Providing metadata for what you post in the new public of the Net enables it to be found more easily. We can also make more sense of it, just as a caption helps us make sense of a photo.

Metadata also helps with the second problem inherent in an open, superabundant system: Most of what's posted will be crap. So, we need ways to evaluate and filter, which can be especially difficult since what is crap for one effort may be gold for another.

Some of this increase in metadata requires explicit effort by humans: We curate collections, we rate items, we leave comments, we write reviews. But much valuable metadata can be deduced by examining the trails we inadvertently leave behind. Sites such as Amazon have become expert at this: By analyzing data about what people click on and what they buy individually and in aggregate, Amazon is able to tempt visitors with other items they might be interested in. The data Amazon uses for this purpose is not left by visitors on purpose. For that very reason, it tells a special truth.

Over the past several years, there has been an effort under way, beginning in geeky quarters, to add metadata to the Net systematically so that data from around the Web can be harvested and used. For example, a local library site, Google Books, and Amazon all might have information online about the book *On the Origin of Species:* bibliographic information, ratings, reviews, analyses of word frequencies, and more. But because each of these has a different way of identifying the book, there's no easy way to write a program that will reliably pull all that information together. If each of these sites followed the conventions specified by the Semantic Web—initiated by Sir Tim Berners-Lee, the inventor of the World Wide Web, around the turn of the millennium—computer programs could far more easily know that these sites were referring to the same book. In fact, the Semantic Web would make it possible to share far more complex information from across multiple sites. Agreeing on how to encode metadata makes the Net capable of expressing more knowledge than was put into it. That is the very definition of a smart network.

But creating that metadata can be difficult, especially since many Semantic Web adherents originally proceeded by trying to write large, complex, logical representations of domains of the world. Writing these ontologies, as they are called, can be difficult. If you're just trying to write a model of, say, knitting, it might not be too complex; you'd have to represent all the objects (needles, yarns, patterns, knitters, knit goods, etc.) and all their relationships (knit goods have knitters, knitters use needles, needles have sizes, etc.). But writing an ontology of financial markets would require agreeing on exactly what the required definitional elements of a "trade," "bond," "regulation," and "report" are—as well as on every detail and every connection with other domains, such as law, economics, and politics.

So, some supporters of the Semantic Web (including Tim Berners-Lee[8]) decided that there would be faster and enormous benefits to making data accessible in standardized but imperfect form—as what is called "Linked Data"—without waiting for agreement about overarching ontologies. So, if you have a store of information about, say, chemical elements, you can make it available on the Web as a series of basic assertions that are called "triples" because they have the form of two objects joined by a relation: "Mercury is an element." "Mercury has an atomic weight of 200.59." "Mercury has a boiling point of 356.73C." "Mercury costs $694.62 per pound."[9] Let all this data be publicly accessible and researchers, developers, and business people will benefit from applications that pull together data from all over the Web to discover new properties, uses, and markets for mercury. Do this for massive amounts of data, from the worlds of science, business, medicine, culture, government, economics, sociology—and we now have an unprecedented resource for discovering new ideas based on what we already know about our world. This resource is a *data commons,* and we are just beginning to understand its transformational effect.

One of the keys to the success of Linked Data is that it isn't too fussy about its metadata. So, should the author of a book be referred to in the metadata about that book as "author," "writer," or "creator"? Rather than having to settle that question, Linked Data advocates will tell you that instead of using any of those terms in your "triples," express the relationship as a link that points at some well-known site that has defined the relationship for you. So, instead of using the word "author," put in a link to what the Dublin Core standard[10]—a vocabulary of terms useful for publishing documents—defines as the authorial relation. Now any application that wants to understand your triple knows that the relationship is the one defined over on the Dublin Core site. This approach may be messy and imperfect, but it is 100 percent better than not releasing data because you haven't figured out how to get the metadata perfectly right.

The rise of Linked Data encapsulates the transformation of knowledge we have explored throughout this book. While the original Semantic Web emphasized building ontologies that are "knowledge representations" of the world, it turns out that if we go straight to unleashing an abundance of linked but imperfect data, making it widely and openly available in standardized form, the Net becomes a dramatically improved infrastructure for knowledge.

Linked Data is nevertheless itself only an example of a more expansive practice: Create metadata so your information can be reused. Linked Data is usable because it points beyond itself to information about the information. That's how a "triple" about mercury can be identified as being about the chemical, the planet, or the Roman god. The atoms of data hook together only because they share metadata.

Indeed, a little metadata can go a very long way. This is important because in the Net of abundance we need more metadata about the authority of works than credentialed institutions can provide. This has been one of the richest areas of innovation since the Web began. For example, a notice that 99.9 percent of an eBay seller's 14,000 transactions have been rated as satisfactory is a better guide than knowing that the seller teaches at Oxford. Knowing that your social network liked a local restaurant is more useful than that the persnickety local restaurant critic did not. We need to keep developing such systems, and there is little doubt that we will.

A Net richer in metadata is richer in more usable and useful knowledge.

3. Link everything

It turns out that it's good to show your work. For example, you could ask WolframAlpha to divide the per capita income of Germany plus Italy by the per capita income of France plus England. If the answer (98.6 percent) does not meet your expectations, WolframAlpha has a "Show details" link that gives you the information it used in doing the calculation.

Of course, not all examples of showing your work are that literal. When journalism blogger Jay Rosen links to a source that supports a statement, he is being transparent about how he came to a conclusion, while also increasing the authority of his work.

Linking also situates your work within its context, tempting us to learn more. When Jillian York blogged a thoughtful response to a Malcolm Gladwell article on social media and the Arab spring, she hyperlinked to a post by Tunisian blogger Sami Ben Gharbia.[11] York refers us there not simply to acknowledge where she got the idea but to encourage us to drop into the web of which Ben Gharbia's blog is the center.

This generous behavior is the opposite of the typical commercial site, at least in the early days of the Web, that tried to keep readers on its site no matter what. And it is also the opposite of much print-based writing that wants to get you from the first page to the last without interruption or distraction. From this has come a new model of what it means to read and study, hyperlinking in and out of an article. Works written for uninterrupted reading certainly can have tremendous value. But a web of ideas that lets us peek behind the page, and that lets us fan out across a territory chased by our interests, is a model of smartness as well.

More links, more links.

4. Leave no institutional knowledge behind

Before there was the Net, institutions such as universities put people into the same space to develop ideas, provided criteria for what counted as knowledge, and generated credentials that let others trust that knowledge.

The strengths of traditional institutions are also their weaknesses. They bring knowers together, but very few applicants make the cut. The isolation that institutions provide also isolates thought. Institutions tend to become echo chambers, even if we call them "schools of thought."

The Net's wild connectivity, on the other hand, doesn't respect the boundaries within which institutions operate. Not only do uncredentialed people outside the institution get to mix it up with those inside of it, those inside the institution cannot keep their work unmixed. The scientist doing experiments at her bench finds herself embroiled in a conversation about the ethical or environmental implications of her work. Then that spawns a personal relationship—perhaps fleeting, perhaps an animosity—that occasionally gives rise to a flurry of political activity. The Net refuses to keep information apart from communication and apart from sociality.

The Net is not going to tear down all institutions. Rather, institutions are already becoming deeply enmeshed in the Net. And the Net is developing its own institutions with—perhaps inevitably—some of the same weaknesses as traditional ones.[12] If the Net is to be our new infrastructure of knowledge, it must take advantage of all the knowledge already developed by our existing institutions.

Some of those institutions are highly reluctant to give up the goods because their economic well-being has been based on them. The *New York Times* has been struggling with this question. Caught between the social responsibility it's given itself as "the paper of record" and its economic responsibilities as a commercial enterprise, it has struck a cumbersome compromise based on the date of publication. So (at least as I write this), you can read the article announcing the fall of the Berlin Wall on November 9, 1989, online and for free.[13] If, however, you want to read E. J. Dionne's article about the presidential campaign from exactly three years before that, the *Times* shows you the first paragraph and offers to sell you access to the rest for $3.95.[14] If you are looking for an article published between 1923 and 1980—perhaps about the twentieth birthday of the "mental hygiene movement" in 1929[15]—the *Times* won't show you even a paragraph, but will sell it to you for $3.95. Finally, if you want to read anything in the *Times*'s database of 13 million articles that was published before 1923, when the chain of copyright extensions gripped us like a communicative disease, you can, for free. The *Times*'s crazy quilt of permissions can be frustrating, but if you are researching within the archives' free years, you may weep tears of joy and gratitude. The Net is made much smarter by the presence of this treasure.

Our existing institutions have produced so much knowledge over the centuries that it would be tragic not to bring all of that knowledge to the Net. For example, we should be encouraging many more colleges to adopt the OpenCourseWare approach championed by the Massachusetts Institute of Technology that puts videos of classes on the Web for free.[16] And libraries not only have content in books and articles, they have the expertise of librarians, they have metadata about usage patterns that can be used to guide researchers, and they are at the center of communities of scholars who are the most learned people in their fields.

The Net becomes systemically smarter when all of this is made available. Not only will individuals find it and use it, but when information is made programmatically accessible to other computers, developers will come up with ways to magnify the value of that information by aggregating it, analyzing it, connecting it, and mashing it up. Traditional institutions need to be fierce contributors to the Net if our new infrastructure is to move us toward knowledge and not dazed ignorance.

5. Teach everyone

Say something foolishly technodeterministic in the presence of Eszter Hargittai and she will produce the data that show that if you just sit people in front of the Net, they will not necessarily become Net enthusiasts, Net experts, or even Net users. Hargittai is a sociologist who loves the Net, but loves data at least as much. Her careful studies have found that success with the Net varies according to the same old bundle of factors around class, income, and education. We need to *teach* everyone how to use the Net.[17]

If we want the Net to move knowledge forward, we should also remember the differences among the world of people who might come to our sites. The Net's openness means that some of our visitors will not know the ethos of the site. Explicit explanations of the type of conversation permitted and the quality of the information posted are therefore very helpful.

But no amount of explicit metadata is going to suffice. The Net is going to remain not just a commons but a wilds. At least we can hope so. If we want the Net to move knowledge forward, then we need to educate our children from the earliest possible age about how to use the Net, how to evaluate knowledge claims, and how to love difference.

This first task is the easiest, although still considerable. Given the complexity and magnitude of the Net, it is remarkably easy to learn how to operate it. But knowing how to click buttons is the least of our concerns.

The second task—learning how to evaluate knowledge claims—is never-ending. Now that the temple priests don't control what we encounter, we need those critical-thinking skills more than ever. The Internet pioneer Howard Rheingold talks about these as "literacies." For example, we need to get better at distinguishing lying crap from well-documented conclusions, becoming more open to new ideas, and learning how to participate in a multi-way, multi-cultural discussion.[18] The journalist Dan Gillmor has been writing about the skills citizens need to make sense of—and participate in—the new media ecology.[19] Ethan Zuckerman has been thinking deeply about our tendency toward a smug homophily (our preference for others like us) and about structural ways we might get ourselves interested in something other than our own echoes.[20] We are just at the beginning of figuring out what behaviors and attitudes lead to a smarter network.

The third task—learning to love difference—is the hardest. As we have seen, echo chambers are a requirement for the discussion and collaboration that advance knowledge, and even echo chambers with solid walls can serve some purposes, such as engendering political enthusiasm. But we also know that we make ourselves stupid when we restrict ourselves to tolerating only the mildest disruptions of our comfort. For the Net to maximize its capacity for knowledge, then, we need to push past our urge to stick with people like us.

Unfortunately, because homophily seems to be a strong human propensity and in some degree a requirement for knowledge itself, no answer will suffice. We can try to counter our tendency to stay safely in our home neighborhood by visiting sites that expand our range. We can link to such sites in order to encourage others. And we can revel in—and recommend to others—the voices of others from around the world, as well as works of literature, journalism, and art that delight us with their presentation of cultural differences. But we will never succeed at being as cosmopolitan as we should be.[21] Life is local. Without the local, we would have no standpoint by which to make sense of the world near us or the world within which the local is embedded.

The Net, however, presents us with both an opportunity and a model.

The opportunity: The Net lowers the barriers to encountering and interacting with that which is different. The barriers that remain are not our technology's but our own. We have lost every excuse not to embrace difference.

The model: We can understand our tendency to be concerned only with what is closest and most familiar to us not in the old geographic way in which we think of ourselves as a point on a map with circle of tight radius drawn around us. Instead, we can understand ourselves as a Web page interpenetrated with links, connected to a world that makes sense of us, and that takes us up and makes us interesting. Since we historically understand ourselves in terms of our technology, perhaps our hyperlinked infrastructure will give us a self-understanding that makes it easier for our curiosity and compassion to overcome our self-centered fears.

—

Those are five areas in which we can work to make the Net a better place for knowledge. And there is reason for hope.

The past fifteen years have proven that we are capable of working together, often for free, building places of knowledge that any sane person would once have dismissed as impossibly ambitious. We have built not just the greatest encyclopedia in human history but also scholarly indexes, open access repositories, competing commons of photographs, a universally accessible "Encyclopedia of Life," maps of galaxies, and annotated genomes. Clay Shirky attributes this burst of collaborative creativity to what he calls a "cognitive surplus"[22] that can now be applied to problems of enormous difficulty, thanks to a scalable network that lets us connect to strangers from around the world who are eager to pitch in.

As many have rightly pointed out, our motives for collaborating are rarely pure. We apply ourselves to these problems for all the motives that humans have. But we've always pursued knowledge for an impure mix of reasons, simply because we are humans. Indeed, the rich variety of our motives is itself reason to hope: We now have more reasons than ever to join in the collective pursuit of knowledge. There is so little standing in our way of learning and contributing that the weakest of reasons can be enough to bring us to contribute. We thus do not yet have any good idea of what *cannot* be done by connected humans when working at the scale of the Net.

The Next Darwin

Charles Darwin is not replaceable.

He made a leap of thought that no one could have predicted. The leap took years of work and thousands of miles of travel, all of which he undertook having no idea that he was preparing for a new idea of such magnitude.

The insight was not deducible simply from the observations he'd made. Had Darwin not been preceded by Lamarck, he might not have thought that evolution was possible at all. Had Darwin not taken with him on the *Beagle* Charles Lyell's *Principles of Geology,* he might not have had the idea that a population could change gradually over an enormous period of time. Further, Darwin developed his insight within a rich network of colleagues and correspondents. Then, when he was at long last ready, Darwin's work changed history because it raced through the nineteenth-century network of scientists, and then of writers, and then throughout the networks we call culture and history.

Today there is a branch of science called environmental niche modeling[23] that takes advantage of the vast databases of information about the global distribution of species and the environments in which they live. For example, a few years ago, a team of scientists modeled the distribution of eleven species of chameleons on the island of Madagascar.[24] They used information gathered in recent surveys and examined specimens in collections in museums. They harvested data from remote sensors, from weather stations, and from geological surveys. They correlated the twenty-five different sorts of data they'd aggregated with where they knew various chameleon species lived, and used that to predict the likelihood of finding closely related species in areas of Madagascar with similar profiles. The model led them to discover seven new species of chameleon.

That is good science, although it is of course not near the level of Darwin's insight. Nevertheless, we should expect that the next Darwin is more likely to be a data wonk than a naturalist wandering through an exotic landscape. Programs like Eureqa will get more and more sophisticated, and may start noticing the correlation of chameleon distribution and environmental factors, or might even "think" it odd that finch beaks vary from island to island in the Galapagos archipelago. The next Darwin will make sense of data that a computer has uncomprehendingly flagged as interestingly anomalous.

The next Darwin is likely to do her work in public, which is to say, on the connected Net. Rather than waiting to publish final results, she will post early results and perhaps a speculative hypothesis. As word gets out, a web of links will grow around her. Some nodes will turn into hubs, at least for a day or a month. There is no predicting whether the owners of those hubs will be professionals or amateurs, scientists or businesspeople, scholars or wags. We can predict, however, that many of the nodes and the threads that connect them will disagree, will argue, will get it wrong, will be childish and egoistic, will be a waste of the digital silk that links them. Nevertheless, we will now see how the idea spreads and the effect it has as the competent and the crazy take it up, make it their own, and pass it on.

This is not just a change in the tools of knowledge. The nature of the knowledge that the next Darwin uncovers will be different from that produced by Charles Darwin a hundred and fifty years ago. Our new knowledge does not consist of a careful set of works that have passed through a series of narrow gates. We thought that knowledge was scarce, when in fact it was just that our shelves were small. Our new knowledge is not even a set of works. It is an infrastructure of connection. We now travel through abundance as knowingly as we can, which is to say always within a context and from a standpoint, always with others, always with the amount of care we judge is required, always fallibly. Knowledge has become a network with the characteristics—for better and for worse—of the Net.

We will argue about whether our new knowledge will bring us closer to the truth, as I think it overall does. But one thing seems clear: Networked knowledge brings us closer to the truth about knowledge.

ACKNOWLEDGMENTS

SO MANY PEOPLE CONTRIBUTED in so many ways to this book that I am doomed to fail in acknowledging them all. It is difficult to thank a loose-edged network.

Two institutions have been especially helpful. Harvard's Berkman Center for Internet & Society where I am a senior researcher, and the Harvard Library Innovation Lab, where I am co-director, alongside Kim Dulin, together provide the intellectual milieu that forms and challenges ideas, the companionship and collegiality that turns tasks into pleasure, and the resources to pursue ideas.

I am especially grateful to the members of the Berkman Book Club (which is for writing books, not reading them) for their critiques as this book developed, and for the emotional support all writers need even when we pretend we're pretending otherwise.

A long list of people have helped me with ideas, by reading sections, by engaging in discussions with me, and in general by being generous with their time and expertise. This list is terribly incomplete, but it is at least alphabetical: AKM Adam, Jacob Albert, Patti Anklam, Solon Barocas, Isabel Walcott Draves, Mark Federman, Dan Gillmor, Timo Hannay, Terry Heaton, Harry Lewis, Maura Marx, Bob Morris, Beth Noveck, Andy Orem, Howard Rheingold, Peter Suber, Barbara Tillett, Catherine White, and John Wilbanks.

I owe a special debt to a few other people. I am lucky beyond words to have access to Ethan Zuckerman's outsized brain and heart, and to his friendship. John Palfrey has supported me as a colleague and friend and, despite my having a few decades on him, has been an important mentor. The team of developers and designers at the Library Innovation Lab every day expands the horizons of what I thought possible. Eszter Hargittai's rigorous research standards and good humor have affected me, albeit never enough. Christian Sandvig has been generous with his ideas and sources over the course of always stimulating discussions. Clay Shirky as a friend and thinker has more influence on me than he thinks or wants. Tim Sullivan gave me permission to let the topic assume the form that it wanted. Tim Bartlett, my editor, gave this book the systematic and insightful critique authors dream of. Christine Arden gave it a superb copy-editing pass that improved it on multiple levels. David Miller, my agent and friend, was essential to getting me from an idea to a book; it has been my tremendous good fortune to have worked with him and Lisa Adams on four books over the past ten years. My wife, Ann Geller, is my first reader and love. Our three children—Nechama, Leah, and Nathan—have lost none of their capacity to delight as they have grown into two women and one man.

All mistakes and errors are solely the responsibility of Wikipedia.

NOTES

Prologue: The Crisis of Knowledge

1. David Barstow, Laura Dodd, James Glanz, Stephanie Saul, and Ian Urbina, "Regulators Failed to Address Risks in Oil Rig Fail-Safe Device," *New York Times,* June 19, 2010, http://www.nytimes.com/2010/06/21/us/21blowout.html.

2. Sam Tanenhaus, "John Updike's Archive: A Great Writer at Work," *New York Times,* June 19, 2010, http://www.nytimes.com/2010/06/21/books/21updike .html.

3. Jeffrey Marcus, "When a Soccer Star Falls, It May Be Great Acting," *New York Times,* June 20, 2010, http://www.nytimes.com/2010/06/21/sports/soccer/ 21diving.html.

4. These "previously unreleased notes" are available online at http:// documents.nytimes.com/documents-on-the-oil-spill?ref=us#document/p34.

5. Kantor made this point at the first Edelman conference on media and public relations in June 2007.

Chapter 1: Knowledge Overload

1. R. L. Ackoff, "From Data to Wisdom," presidential address to ISGSR in June 1988, *Journal of Applied Systems Analysis* 16 (1989): 3–9.

2. See Milan Zeleny, "Management Support Systems: Towards Integrated Knowledge Management," *Human Systems Management* 7 (1987): 59–70; Michael Cooley, *Architecture or Bee?* (Hogarth Press [London], 1987), mentioned in Nikhil Sharma, "The Origin of the 'Data Information Knowledge Wisdom' Hierarchy," February 4, 2008, http://nsharma.people.si.umich.edu//dikw_origin.htm; and Harlan Cleveland, "Information as Resource," *The Futurist,* December 1982, pp. 34–39, also in Sharma.

3. The history of this concept is summarized in Nikhil Sharma, "The Origin of the 'Data Information Knowledge Wisdom' Hierarchy," February 4, 2008, http://nsharma.people.si.umich.edu//dikw_origin.htm. For more on Ackoff, see Gene Bellinger, Durval Castro, and Anthony Mills, "Data, Information, Knowledge, and Wisdom," 2004, http://www.systems-thinking.org/dikw/dikw.htm.

4. "The IBM 650" (part of an online history at the IBM site), http://www-03.ibm.com/ibm/history/exhibits/650/650_intro.html.

5. Leonard Dudley, *Information Revolution in the History of the West* (Edward Elgar Publishing, 2008), p. 266.

6. The transformation of "information into instructions" is discussed on pages 163–180 of Jennifer Rowley's "The Wisdom Hierarchy: Representations of the DIKW Hierarchy," *Journal of Information Science* 33 (February 14, 2007), DOI: 10.1177/0165551506070706. This article provides an excellent roundup of the various forms of the data-to-wisdom hierarchy.

7. Russell L. Ackoff, *Re-creating the Corporation: A Design of Organizations for the 21st Century* (Oxford University Press, 1999), p. 160.

8. Skip Walter, "Knowledge vs. Information," Extreme Productivity by Design blog, January 2, 2008, http://factor10x.blogspot.com/2008/01/knowledge-versus-information.html.

9. Zeleny, "Management Support Systems," p. 59.

10. Frank E. Smitha, "An Imperfect Democracy," in *Macrohistory and World Report,* http://www.fsmitha.com/h1/hello4.htm.

11. The IBM 650 could use the "IBM 650 Magnetic Drum Data Processing machine with a series of disk memory units, which are capable of storing a total of 24-million digits" (http://www-03.ibm.com/ibm/history/exhibits/650/650_pr2.html). I am assuming that a desktop computer these days has a terabyte of hard-disk space.

12. Alvin Toffler, *Future Shock* (Random House, 1970), p. 350. The term "information overload" appeared as early as 1962; see Bertram M. Gross, "Operation Basic: The Retrieval of Wasted Knowledge," *Journal of Communication* 12 (1967): 67–83, DOI: 10.1111. And Norbert Wiener talked about overloading the nervous system even earlier in his 1948 book *Cybernetics* (MIT Press, reprinted in 1961).

13. The concept of sensory overload was itself new. It's often traced back to an article by Georg Simmel, written in 1903, that explained how the overwhelming sensations experienced by city-dwellers can make them reserved and unresponsive. The term "sensory overload" doesn't show up until the 1950s, and hit public consciousness in the late 1960s only when it became useful for warning kids to stay away from psychedelics.

14. Toffler, *Future Shock,* p. 301.

15. Ibid.

16. Ibid.

17. Writing ten years later, one of the authors, Jacob Jacoby, criticized his own research: "Respondents were told that each of the brands of rice/prepared dinners were either high or low in calories per serving." Looking back, Jacoby commented that this was too artificial a restriction: "In contrast, information confronting the consumer in the real world is more complex." See Jacob Jacoby, "Perspectives on Information Overload," *Journal of Consumer Research* (March 1984): 432–435 at 432.

18. Richard Saul Wurman, *Information Anxiety* (Doubleday, 1989), p. 35, citing Peter Large, *The Micro Revolution Revisited* (F. Pinter, 1984).

19. Wurman, *Information Anxiety*, p. 34.

20. Roger E. Bohn and James E. Short, "How Much Information? 2009 Report on American Consumers," Global Information Industry Center, University of California–San Diego (2009), p. 7, http://hmi.ucsd.edu/howmuchinfo.php.

21. This figure is cited at http://en.wikipedia.org/wiki/Zettabyte.

22. Quoted in Ann Blair, "Reading Strategies for Coping with Information Overload ca. 1550–1700," *Journal of the History of Ideas* 63, no. 1 (January 2003): 11–28 at 15.

23. Quoted in Daniel Rosenberg, "Early Modern Information Overload," *Journal of the History of Ideas* 63, no. 1 (January 2003): 1–9. at 1, http://www.jstor.org/stable/3654292.

24. Quoted in Richard I. Yeo, "A Solution to the Multitude of Books: Ephraim Chambers's 'Cyclopedia' (1728) as 'The Best Book in the Universe,'" *Journal of the History of Ideas* 63, no. 1 (January 2003): 61–72 at 62.

25. Lucius Annaeus Seneca, *Dialogues and Letters,* translated by Charles Desmond Nuttall Costa (Penguin, 1997), p. 45.

26. Quoted in Yeo, "A Solution to the Multitude of Books," p. 62.

27. Bohn and Short, "How Much Information?" p. 7; (2007). The reference to a prior report of 0.3 zettabytes also comes from this source.

28. Clay Shirky, keynote address at Web2.0 Expo, September 16–19, 2010, http://web2expo.blip.tv/file/1277460/.

29. Mary Spiro, *Baltimore Science News Examiner* blog, July 20, 2009, http://www.examiner.com/x-6378-Baltimore-Science-News-Examiner~y2009m7d20-Science-on-the-airwaves-eight-podcasts-you-shouldnt-miss.

30. Between 1900 and 1909, 83,512 new books were published in the United States. See Jacob Epstein, *The Book Publishing Industry* (W. W. Norton, 2002), p. 21; and Bowker press release, May 19, 2009, http://www.bowker.com/index.php/press-releases/563.

31. Google gives many more hits if you omit the quotation marks around the phrase.

32. Beth Simone Noveck, *Wiki Government* (Brookings Institution Press, 2009).

33. Gartner 2008 Annual Report, https://materials.proxyvote.com/Approved /366651/20090408/CMBO_39013.

34. Interview with Jack Hidary, May 21, 2009.

35. Interviews with Beth Noveck (February 2010) and the head of Expert Labs, Anil Dash (December 2009 and January 2010). I was a participant in the organizing meeting.

Chapter 2: Bottomless Knowledge

1. Bennett Cerf, *Try and Stop Me* (Simon & Schuster, 1944), p. 75.

2. "U.S. Census Bureau's Budget Estimates as Presented to Congress: Fiscal Year 2010," May 2009, http://www.osec.doc.gov/bmi/budget/10CJ/Census% 2520FY%25202010%2520Congressional.pdf.

3. Alvin Powell, "John Enders' Breakthrough Led to Polio Vaccine," *Harvard University Gazette,* October 7, 1998, http://www.news.harvard.edu/gazette /1998/10.08/JohnEndersBreak.html.

4. Bill Clinton, "How We Ended Welfare, Together," *New York Times,* August 22, 2006, http://www.nytimes.com/2006/08/22/opinion/22clinton.html.

5. Paul Rosenberg, "The Myth That Conservative Welfare Reform Worked— Part 1," OpenLeft.com, February 27, 2010, http://openleft.com/diary/17541/the -myth-that-conservative-welfare-reform-workedpart-1.

6. Jason Deparle and Robert M. Gebeloff, "Living on Nothing but Food Stamps," *New York Times,* January 2, 2010, http://www.nytimes.com/2010/01 /03/us/03foodstamps.html?_r=1&th&emc=th.

7. Quoted in Samuel A. Kydd, *The History of the Factory Movement from the Year 1802 to the Enactment of the Ten Hours' Bill in 1847, Volumes 1–2* (Ayer Company Publishers [London], 1857), pp. 46–47, http://books.google.com /books?id=qdTscxnOXfoC.

8. Quoted in ibid., p. 51.

9. This information is available at http://www.etymonline.com/index.php ?term=fact.

10. J. A. Simpson, *Oxford English Dictionary* (Oxford University Press, 2009).

11. Mary Poovey, *A History of the Modern Fact* (University of Chicago, 1998).

12. The title of the work in which Bacon lays this out—*Novum Organum*—is a reference to Aristotle's *Organum.*

13. See Chapter 3 of Barry Gower's *Scientific Method* (Routledge, 1997), http://books.google.com/books?id=D3rV2t2XkWYC&pg=PA40&lpg=PA40.

14. Ibid., p. 49.

15. Poovey traces the role of interests to Hobbes and following thinkers.

16. Thomas Malthus, *An Essay on the Principle of Population,* Vol. 1 (first edition). This is online at http://www.econlib.org/library/Malthus/malPop1.html.

17. Malthus, *An Essay on the Principle of Population,* p. 229.

18. "Chimney Sweepers' Regulation Bill," *Hansard* 39 (February 16, 1819): 448–454, http://hansard.millbanksystems.com/commons/1819/feb/17/chimney -sweepers-regulation-bill.

19. Sir Llewellyn Woodward, *The Age of Reform 1815–1870,* 2nd ed. (Clarendon Press, 1962), p. 36.

20. Michael J. Cullen, *The Statistical Movement in Early Victorian Britain: The Foundations of Empirical Social Research* (Barnes & Noble, 1975), pp.10ff.

21. In fact, the emblem of the *Journal of the Royal Statistical Society* was initially grain waiting to be threshed by others; see ibid., p. 47.

22. Charles Kingsley's *Yeast* and Disraeli's *Sybil* are both based on the same blue book; see Roger P. Wallins, "Victorian Periodicals and the Emerging Social Conscience," *Victorian Periodicals Newsletter* 8 (June 1975): 29, 47–59.

23. Charles Dickens, *Hard Times* (Hurd and Houghton, 1870), p. 14, http:// books.google.com/books?id=ORgKTQ66LN4C.

24. Ibid., p. 15.

25. Ibid., p. 14.

26. Ibid., p. 130.

27. Kenneth Bensen, *Charles Dickens: The Life of the Author* (New York Public Library, 2002), http://www.fathom.com/course/21701768/session1.html.

28. The arbitration transcript is available at http://www.alohaquest.com /arbitration/transcript_001208.htm. See also Lili'uokalani's protest from Blount's report at http://libweb.hawaii.edu/digicoll/annexation/protest/liliu2.html.

29. Edwin Brown Frimage, "Fact-Finding in the Resolution of International Disputes—From the Hague Peace Conference to the United Nations," *Utah Law Review* (1971): 421–473, http://content.lib.utah.edu/u?/ir-main,8725; Thomas M. Franck and Laurence D. Cherkis, "The Problem of Fact-Finding in International Disputes," *Western Reserve Law Review* 18 (1966–1967): 1483–1524.

30. For further discussion of this activity, see the index of the *New York Times.*

31. Henry David Thoreau, entry on April 19, 1852, in *The Journal of Henry D. Thoreau 1837–1855,* Vol. 1 (Dover Publications, 1962), p. 384.

32. Cited by John Updike in his introduction to Henry David Thoreau's *Walden* (Princeton University Press, 2004), p. xiii.

33. Letter to J. D. Hooker, May 10, 1848, in *More letters of Charles Darwin: A Record of His Work in a Series of Hitherto Unpublished Letters,* Vol. 1 (J. Murray, 1903), p. 65.

34. See Jeffrey Cramer's introduction to Thoreau in Jeffrey S. Cramer, ed., *Walden: A Fully Annotated Edition* (Yale University Press, 2004), p. xx.

35. Charles Darwin, *A Monograph of the Sub-Class Cirripedia, with Figures of All the Species. The Lepadidæ; or,Pedunculated Cirripedes,* Vol. 1 (The Ray Society, 1852), p. 77, http://darwin-online.org.uk/content/frameset?itemID=F339.1&viewtype=side&pageseq=1.

36. White House Press Office, "Transparency and Open Government," January 21, 2009, http://www.whitehouse.gov/the_press_office/transparencyandopengovernment/.

37. Vivek Kundra, "They Gave Us the Beatles, We Gave Them Data.gov," January 21, 2010, http://www.whitehouse.gov/blog/2010/01/21/they-gave-us-beatles-we-gave-then-datagov.

38. From a talk by Beth Noveck at Harvard Law School on April 29, 2010, http://www.hyperorg.com/blogger/2010/04/29/berkman-2b2k-beth-noveck-on-white-house-open-government-initiatives/.

39. Clay Shirky provides an excellent explanation of the importance of defaults in *Cognitive Surplus: Creativity and Generosity in a Connected Age* (Penguin, 2010).

40. This information can be found at http://www.fueleconomy.gov/feg/download.shtml.

41. Lawrence Lessig, "Against Transparency," *The New Republic,* October 9, 2009, http://www.tnr.com/article/books-and-arts/against-transparency.

42. Mike Scannell, an engineer at Ford Motor, figures that a typical 90-row punchcard encodes 112.5 bytes and that it would therefore take 9,544,472 punchcards to encode a gigabyte. See "Re: How Many IBM Punch Cards in 20 Giga Bytes of Data?", *MadSci,* August 26, 2000, http://www.madsci.org/posts/archives/2000–08/967332303.Cs.r.html.

43. Scott M. Nelson and Debbie Lawlor, "Predicting Live Birth, Preterm Delivery, and Low Birth Weight in Infants Born from In Vitro Fertilisation: A Prospective Study of 144,018 Treatment Cycles," *Public Library of Science Medicine* 8, no. 1 (January 2011), DOI: 10.1371/journal.pmed.1000386, http://www.plosmedicine.org/article/info%3Adoi%2F10.1371%2Fjournal.pmed.1000386.

44. See http://www.hfea.gov.uk/fertility-treatment-facts.html.

45. See the Linked Data Web site at http://www.linkeddata.org.

46. See http://en.wikiquote.org/wiki/Daniel_Patrick_Moynihan.

Chapter 4: The Expertise of Clouds

1. *Report of the Presidential Commission on the Space Shuttle Challenger Accident,* 1986, http://www.chron.com/content/interactive/special/challenger/docs/report.html.

2. Geoffrey Chaucer, *The Complete Works of Geoffrey Chaucer: Boethius and Troilus,* edited by Walter William Skeat (Clarendon Press, 1900), http://books .google.com/books?id=xWERAAAAIAAJ.

3. Donald Abelson, *A Capitol Idea: Think Tanks and U.S. Foreign Policy* (McGill-Queen's Press, 2006), p. 50. Note that Abelson disagrees with this dating of the origins by Dickson.

4. James A. Smith, *The Idea Brokers* (Free Press, 1991), pp. 24–27. See also "Why Think Tanks," *United Press International,* January 10, 2001, as well as Albert Wiggam's *The New Decalogue,* cited by Thomas Leonard in "American Progressives and the Rise of Expertocracy," *History of Economics Society Meetings* (June 2006), Grinnell, Iowa. Wiggam's book puts forward racist eugenics as a new religion we should all embrace. There's more about Wiggam in Christine Rosen, *Preaching Eugenics: Religious Leaders and the American Eugenics Movement* (Oxford University Press, 2004), pp. 128ff.

5. Leonard, "American Progressives and the Rise of Expertocracy."

6. Ellen Swallow Richards quoted in Leonard, "American Progressives and the Rise of Expertocracy," p. 10.

7. Ibid.

8. Jane Rankin, *Parenting Experts: Their Advice, The Research and Getting It Right* (Praeger, 2005). This book is discussed in Wendy Leopold's "Do Spock, Other Parenting Experts Get It Right?" *Observer Online,* January 26, 2006, http://www.northwestern.edu/observer/issues/2006/01/26/parenting.html.

9. David Riesman, Nathan Glazer, and Reuel Denney, *The Lonely Crowd: A Study of the Changing American Character* (Yale University Press, 1950).

10. The 233 children inside fled. See Leslie M. Harris, *In the Shadow of Slavery: African Americans in New York City, 1626–1863* (University of Chicago Press: Chicago, 2004). An excerpt of the chapter titled "The New York City Draft Riots of 1863" can be found at http://www.press.uchicago.edu/Misc /Chicago/317749.html.

11. Howard Rheingold, *Smart Mobs: The Next Social Revolution* (Basic Books, 2003).

12. James Surowiecki, *The Wisdom of Crowds* (Random House, 2004).

13. Jeff Howe, "The Rise of Crowdsourcing," *Wired* 14, no. 6 (June 2006), http://www.wired.com/wired/archive/14.06/crowds.html. See also Howe's *Crowdsourcing* (Crown Business, 2008).

14. "Darpa Network Challenge: We Have a Winner," https://network-challenge.darpa.mil/Default.aspx.

15. "How It Works" (MIT), 2009, http://balloon.mit.edu/mit/payoff/.

16. Darren Murph, "MIT-Based Team Wins DARPA's Red Balloon Challenge, Demonstrates Power of Social Networks (and Cold Hard Cash)," December 6,

2009, http://www.engadget.com/2009/12/06/mit-based-team-wins-darpas-red
-balloon-challenge-demonstrates/.

17. See Jonathan Zittrain's "Minds for Sale" video at http://www.youtube.com
/watch?v=Dw3h-rae3uo.

18. My book *Everything Is Miscellaneous: The Power of the New Digital Disor-
der* (Times Books, 2007) discusses this issue at length.

19. Calvin Trillin, "Where's Chang?" *The New Yorker,* March 1, 2010, pp. 26–29.

20. See the Office of Response and Restoration home page at http://
response.restoration.noaa.gov/type_subtopic_entry.php?RECORD_KEY(entry_s
ubtopic_type)=entry_id,subtopic_id,type_id&entry_id(entry_subtopic_type)=6
87&subtopic_id(entry_subtopic_type)=8&type_id(entry_subtopic_type)=4.
(The ORR is part of the National Oceanic and Atmospheric Administration.)

21. Kermit Pattison, "Crowdsourcing Innovation: Q&A with Dwayne Spradlin
of InnoCentive," *FastCompany,* December 15, 2008, http://www.fastcompany
.com/blog/kermit-pattison/fast-talk/millions-eyes-prize-qa-dwayne-spradlin
-innocentive.

22. "InnoCentive Solver Develops Solution to Help Clean Up Remaining Oil
from the 1989 Exxon Valdez Disaster," InnoCentive press release, November 14,
2007, http://www.marketwire.com/press-release/InnoCentive-Solver-Develops
-Solution-Help-Clean-Up-Remaining-Oil-From-1989-Exxon-Valdez-792827
.htm.

23. "2010 International Contest on LTPP Data Analysis," http://www.fhwa
.dot.gov/pavement/ltpp/contest.cfm.

24. Stefanie Olsen, "DOT Proposes Contest to 'Green' Jet Fuel Industry," July
10, 2008, http://news.cnet.com/8301-11128_3-9987675-54.html.

25. Adam Ash, "Deep Thoughts: The Internet, Is It a Stupid Hive Mind, or the
Potential Savior of Mankind?" May 31, 2006, http://adamash.blogspot.com
/2006/05/deep-thoughts-internet-is-it-stupid.html. See also Don Tapscott and
Anthony D. Williams, "Ideagora, a Marketplace for Minds," *BusinessWeek,* February 15, 2007.

26. Cornelia Dean, "If You Have a Problem, Ask Everyone," *New York Times,*
July 22, 2008, http://www.nytimes.com/2008/07/22/science/22inno.html.

27. The British National Maritime Museum has an excellent article on Harrison, by J. O'Donnell (November 15, 2002). See http://www.nmm.ac.uk/harrison.

28. Quoted in Dean, "If You Have a Problem, Ask Everyone."

29. From the Netflix discussion board, September 18, 2009, http://www
.netflixprize.com//community/viewtopic.php?id=1537. Netflix was going to hold
a second round, but researchers discovered that the data it was exposing could be
used to identify Netflix users and their movie choices. See Ryan Singel, "Netflix
Cancels Recommendation Contest After Privacy Lawsuit," Wired.com, March 12,
2010, http://www.wired.com/threatlevel/2010/03/netflix-cancels-contest/.

30. David Pogue, "The Twitter Experiment," *New York Times,* January 29, 2009, http://www.nytimes.com/2009/01/29/technology/personaltech/29pogue -email.html.

31. David Pogue, tweet on January 31, 2011, http://twitoaster.com/country-us/pogue/anybody-know-of-exact-spots-in-manhattan-where-an-att-iphone-call -drops-every-time-for-sure-thanks-in-advance/comment-page-2/.

32. Interview with Mike Wing, vice president of Strategic and Executive Communications at IBM, June 12, 2009.

33. Email from Richard Polt, March 17, 2010.

34. See http://www.facebook.com/group.php?gid=104873080692&v=info.

35. Carlin Romano, "Heil, Heidegger!" *The Chronicle of Higher Education,* October 18, 2009, http://chronicle.com/article/Heil-Heidegger-/48806/.

36. Interview with Michal Cenkl, January 23, 2010.

37. Interview with Jean Tatalias, January 23, 2010.

38. Interview with Les Holtzblatt, January 23, 2010.

Chapter 5: A Marketplace of Echoes?

1. David Halberstam, *The Best and the Brightest* (Random House, 1969).

2. Victor Navasky, "How We Got into the Messiest War in Our History," *New York Times Book Review* (November 12, 1972), http://www.nytimes.com/books /98/03/15/home/halberstam-best.html.

3. Interview with Beth Noveck, February 6, 2010.

4. Scott Page, *The Difference* (Princeton University Press, 2008), pp.137, 158.

5. Ibid., p. 137.

6. Ibid., p. 153

7. Ibid., pp. 159–162.

8. See Noel Sheppard, "Media Slowly Noticing Sotomayor's 'Wise Woman' Comments," *NewsBusters,* June 6, 2009, http://newsbusters.org/blogs/noel-shep pard/2009/06/06/media-slowly-noticing-sotomayors-wise-woman-comments.

9. From the "mini-biography" on Annie Sprinkle's home page: http://www .anniesprinkle.org/html/about/short_bio.html.

10. Email to the author, May 12, 2010.

11. Annie Sprinkle, "My Conversation with an Anti-Porn Feminist," http:// anniesprinkle.org/writings/pocketporn.html. See also Annie Sprinkle and Gabrielle Cody, *Hardcore from the Heart: The Pleasures, Profits and Politics of Sex in Performance* (Continuum International Publishing Group, 2001), pp. 110–118.

12. Interview with Jon Lebkowsky, May 9, 2010.

13. Rheingold made this point as a member of a panel I moderated at the Personal Democracy Forum, June 3, 2010, in New York City, and in conversation afterward.

14. "Where Is the Wikitorial?" *Los Angeles Times,* June 19, 2005, http://www
.latimes.com/news/opinion/editorials/la-wiki-splash,0,1349109.htmlstory.

15. "A Wiki for Your Thoughts." *Los Angeles Times,* June 17, 2005, http://
articles.latimes.com/2005/jun/17/opinion/ed-wiki17.

16. James Rainey, "'Wikitorial' Pulled Due to Vandalism," *Los Angeles Times,*
June 21, 2005, http://articles.latimes.com/2005/jun/21/nation/na-wiki21.

17. "Los Angeles Times Launches Editorial Wiki," *Wikinews,* June 19, 2005,
http://en.wikinews.org/wiki/Los_Angeles_Times_launches_editorial_wiki.

18. Ross Mayfield, "Wikitorial Fork," *Corante* blog, June18, 2005, http://
many.corante.com/archives/2005/06/18/wikitorial_fork.php.

19. Don Singleton, in his blog: "Write the News Yourself," June 20, 2005,
http://donsingleton.blogspot.com/2005/06/write-news-yourself.html.

20. Harvard Law School announcement, http://www.law.harvard.edu/news
/2008/02/19_sunstein.html.

21. Cass Sunstein, *Republic.com* (Princeton University Press, 2001).

22. Ibid., p. 57.

23. Ibid.

24. Ibid., p. 60.

25. Ibid., pp. 65ff.

26. Ibid., p. 69.

27. Ibid., p. 71.

28. Ibid.

29. Ibid.

30. Cass Sunstein, *Republic.com* (Princeton University Press, 2002), p. 206.
Here, Sunstein is referring in general to the question of whether the Internet
poses a threat to democracy.

31. Interview with Clay Shirky, March 30, 2010.

32. See Francesca Polletta, Pang Ching Bobby Chen, and Christopher Anderson, "Is Information Good for Deliberation? Link-Posting in an Online Forum,"
Journal of Public Deliberation 5, no. 1 (2009), http://services.bepress.com/jpd
/vol5/iss1/art2/. In their summary the authors point out that "URL-link posting
not only generated more interaction than did opinions posted without links but
it also responded to what we call the scale and uptake problems of public deliberation. On the negative side of the ledger, far from equalizing deliberation, the
availability of online information may have given additional advantages to already advantaged groups. This was true even in groups that were actively facili-
tated. The availability of online information may also have fostered discussions,
in some instances, that were more opinionated than informed. Information in
the Internet age is newly accessible, we conclude, but is also politicized in unfa-
miliar ways."

33. Matthew Gentzkow and Jesse M. Shapiro, "Ideological Segregation Online and Offline," *National Bureau of Economic Research* (April 2010), http://www.nber.org/papers/w15916. NBER Working Paper No. 15916. See also David Brooks, "Riders on the Storm." *New York Times,* April 19, 2010, http://www.nytimes.com/2010/04/20/opinion/20brooks.html.

34. Gentzkow and Shapiro, "Ideological Segregation Online and Offline," p. 4.

35. Ethan Zuckerman, in his blog: "The Partisan Internet and the Wider World," May 24, 2010, http://www.ethanzuckerman.com/blog/2010/05/24/the-partisan-internet-and-the-wider-world/.

36. Nicholas Carr, *The Shallows: What the Internet Is Doing to Our Brains* (W. W. Norton, 2010), p. 14.

37. Ibid., p. 139, quoting an article by Patricia Greenfield, a "leading developmental psychologist who teaches at UCLA."

38. Carr, *The Shallows,* p. 139.

39. Al Gore, *The Assault on Reason* (Penguin, 2008).

40. Ibid., p. 1.

41. Ibid., p. 6. See also pp. 259 and 260.

42. Ibid., p. 260.

43. Sam Stein, "New GOP Initiative Stumbles Early, Poster Calls For Repealing Civil Rights Act," *Huffington Post,* May 25, 2010, http://www.huffingtonpost.com/2010/05/25/new-gop-initiative-stumbl_n_588748.html.

Chapter 6: Long Form, Web Form

1. This version is valid, unlike Nabokov's parody in his novel *Pale Fire:* "[O]ther men die; but I/Am not another; therefore I'll not die." See *Pale Fire* (Random House, 1989), p. 40.

2. Robert Darnton, *The Case for Books: Past, Present, and Future* (PublicAffairs, 2010), pp. 75–76. This title originally appeared in the *New York Review of Books* (March 18, 1999) as "The New Age of the Book." The version Darnton reprints in *The Case for Books* has passages that he has edited to remove redundancies: http://www.nybooks.com/articles/archives/1999/mar/18/the-new-age-of-the-book/.

3. Ibid.

4. Ibid., p. 77.

5. Ibid., p. 68.

6. Disclosure: I work in the library system that Robert Darnton heads, but I do not report to him even indirectly.

7. Nicholas Carr, *The Shallows: What the Internet Is Doing to Our Brains* (W. W. Norton, 2010), pp. 164, 174.

8. Sven Birkerts, *The Gutenberg Elegies: The Fate of Reading in an Electronic Age* (Faber and Faber, 1994).

9. Sven Birkerts, "Resisting the Kindle," *The Atlantic,* March 2009, http://www.theatlantic.com/magazine/archive/2009/03/resisting-the-kindle/7345/.

10. Birkerts, *The Gutenberg Elegies,* p. 137.

11. Carr, *The Shallows,* p. 196.

12. Ibid., p. 197. Carr's idea that his neurons sprang "back to life," by the way, seems to dispute his book's claim that the ways the Net changes our physical brain cannot easily be undone.

13. See Lewis Hyde's *Common as Air: Revolution, Art, and Ownership* (Macmillan, 2010) for an excellent history of cultural commons that makes this point, focusing on the American Revolutionary period.

14. See http://www.edge.org/discourse/carr_google.html.

15. Email from Jay Rosen, July 17, 2010.

16. See http://journalism.nyu.edu/pubzone/weblogs/pressthink/2010/07/07/obj_persuasion.html.

17. Ibid.

18. Email from Jay Rosen, July 17, 2010.

19. As Mark Federman pointed out on my blog post about this; see http://www.hyperorg.com/blogger/2010/07/18/jay-rosen-carpenter/.

20. David T. Z. Mindich, *Just the Facts: How "Objectivity" Came to Define American Journalism* (New York University Press, 1998), pp. 4–5.

21. I wrote about this on my blog: http://www.hyperorg.com/blogger/2004/07/28/objective-rhetoric/.

22. Glen Johnson, "Kennedy Leads the Attack," *Boston Globe,* July 28, 2004, http://www.boston.com/news/politics/conventions/articles/2004/07/28/kennedy_leads_the_attack/.

23. David S. Broder, "Democrats Focus on Healing Divisions; Addressing Convention, Newcomers Set Themes," *Washington Post,* July 27, 2004, http://www.washingtonpost.com/wp-dyn/content/article/2008/08/13/AR2008081303419.html.

24. Some philosophers would argue that objectivity is a misguided aspiration because all understanding is situated in a particular historical, cultural, linguistic, and psychological standpoint; if we could see the world truly objectively, we wouldn't be able to make any sense of it.

25. Jay Rosen, "The View from Nowhere," PressThink.org, September 18, 2003, http://archive.pressthink.org/2003/09/18/jennings.html. Thomas Nagel's book *The View from Nowhere* was published in 1986 by Oxford University Press.

26. Anthologized in *The American Magazine,* Vol. 6 (1887), p. 112 (The American Magazine Company) (http://books.google.com/books?id=mZ48

AAAAYAAJ), and cited by John Beatty in his review of "Fair and Balanced: A History of Journalistic Objectivity," *Journalism and Mass Communication Quarterly: Association for Education in Journalism and Mass Communications* 83, no. 1 (April 2006), http://vlex.com/vid/fair-balanced-journalistic-objectivity-61539404. The historian Sheila McIntyre links the concern for fairness and accuracy back to the role of ministers in the seventeenth century as reporters and distributors of news; see Sheila McIntyre, "'I Heare It So Variously Reported': News-Letters, Newspapers, and the Ministerial Network in New England, 1670–1730," *New England Quarterly* 71, no. 4 (December 1998), pp. 593–614, http://www.jstor.org/stable/366604.

27. The obvious source for the New Journalism is *The New Journalism*, edited by Tom Wolfe and Edward Warren Johnson (Harper & Row, 1973).

28. Jay Rosen, "Questions and Answers About PressThink," http://journalism.nyu.edu/pubzone/weblogs/pressthink/2004/04/29/q_and_a.html.

29. Malcom Gladwell, "Small Change: Why the Revolution Will Not Be Tweeted," *New Yorker,* October 4, 2010, http://www.newyorker.com/reporting/2010/10/04/101004fa_fact_gladwell. I replied in a blogpost: http://www.hyperorg.com/blogger/2010/10/02/gladwell-discovers-it-takes-more-than-140-characters-to-overturn-a-government/.

30. Malcolm Gladwell, "Does Egypt Need Twitter?" NewYorker.com, February 2, 2011, http://www.newyorker.com/online/blogs/newsdesk/2011/02/does-egypt-need-twitter.html. I posted a reply: http://www.hyperorg.com/blogger/2011/02/04/gladwell-proves-too-much/.

31. Louis Menand, "Books as Bombs," *New Yorker,* January 24, 2011, http://www.newyorker.com/arts/critics/books/2011/01/24/110124crbo_books_menand.

32. For example, South Carolina's celebration of the 150th anniversary of the start of the Civil War set off a new round in the controversy. See Wayne Washington, "150 Years Later, S. Carolina Celebration Sparks New Civil War," McClatchy.com, December 19, 2010, http://www.mcclatchydc.com/2010/12/19/105532/150-years-later-s-carolina-celebration.html.

33. Ted Nelson coined the term "intertwingularity" in *Computer Lib: Dream Machines* (1974). Frank Hecker read my use of the word in *Everything Is Miscellaneous* and tracked down the exact source of Nelson's phrase "Everything is deeply intertwingled," which is harder than it seems because of the nonstandard ways in which Nelson published his work. See details at http://www.everythingismiscellaneous.com/2007/06/09/untwingling-nelsons-intertwingularity-quote/.

34. See WolframAlpha's Frequently Asked Questions at http://www.wolframalpha.com/faqs.html.

Chapter 7: Too Much Science

1. Allison Aubrey, "Nervous About Alzheimer's? Coffee May Help," *Morning Edition,* June 28, 2010, http://www.npr.org/templates/story/story.php?storyId =128110552.

2. Alix Spiegel, "'Mozart Effect' Was Just What We Wanted to Hear," *Morning Edition,* June 28, 2010, http://www.npr.org/templates/story/story.php?storyId =128104580.

3. "Evolution of the National Weather Service," NOAA's National Weather Service, http://www.weather.gov/pa/history/timeline.php.

4. Larry Greenemeier, "Rainforest Climate Change Sensor Station Goes Wi-Fi," *Scientific American,* March 20, 2008.

5. Bernard K. Forscher, "Chaos in the Brickyard," letters section of *Science,* October 18, 1963, p. 339. Thanks to Christian Sandvig for pointing me at this.

6. See http://www.sdss.org/.

7. This figure refers to bases. See Christopher Southan and Graham Cameron, "Beyond the Tsunami: Developing the Infrastructure to Deal with Life Sciences Data," in Tony Hey, Stewart Tansley, and Kristin Tolle, eds., *The Fourth Paradigm: Data-Intensive Scientific Discovery* (Microsoft Research, 2009), p. 117. http://research.microsoft.com/en-us/collaboration/fourthparadigm/.

8. Library of Congress figure confirmed by Steve Herman, at the Library, in an email to the author, February 28, 2011. Thank you to Barbara Tillett, also at the Library, for her help.

9. See http://www.trancheproject.org.

10. Interview with John Wilbanks.

11. Southan and Cameron, "Beyond the Tsunami," p. 117–118.

12. Hiroaki Kitano, "Systems Biology: A Brief Overview," *Science* 295, no. 5560 (March 1, 2002): 1662–1664.

13. For a superb introduction, see Steven Johnson, *Emergence* (Scribner, 2001).

14. See http://www.icosystem.com/labsdemos/the-game/.

15. Eric Bonabeau, "Agent-Based Modeling: Methods and Techniques for Simulating Human Systems," *Proceedings of the National Academy of Sciences* 99, suppl. 3 (May 14, 2002): 7280–7287, www.pnas.org/cgi/doi /10.1073/pnas.082080899.

16. Kellan Davidson, "Eureqa and Technological Singularity," *Ithaca Action News,* May 12, 2010, http://ithacaactionnews.wordpress.com/2010/05/12/ eureqa-and-technological-singularity/.

17. Quoted in Brandon Keim, "Download Your Own Robot Scientist," *Wired Science,* December 3, 2009, http://www.wired.com/wiredscience/2009/12 /download-robot-scientist/#ixzz0vrPoI4G9. See also the exceptional *RadioLab*

program on this topic: "Limits of Science," April 16, 2010, http://www.wnyc.org/shows/radiolab/episodes/2010/04/16/segments/149570.

18. Nicholas Taleb Nassim, *The Black Swan* (Random House, 2007).

19. The story may be apocryphal, according to a report by Nicholas Wade in "A Family Feud over Mendel's Manuscript on the Laws of Heredity," May 31, 2010, http://philosophyofscienceportal.blogspot.com/2010/06/gregor-mendel-and-pea-breeding.html.

20. Jennifer Laing, "Comet Hunter," *Universe Today,* December 11, 2001, http://www.universetoday.com/html/articles/2001–1211a.html.

21. Jennifer Ouellette, "Astronomy's Amateurs a Boon for Science," *Discovery News,* September 20, 2010, http://news.discovery.com/space/astronomys-amateurs-a-boon-for-science.html.

22. Mark Frauenfelder, "The Return of Amateur Science," *Boing Boing,* December 22, 2008, http://www.good.is/post/the-return-of-amateur-science/.

23. Thanks to the people who responded to the request for examples I posted on my Web site: Garrett Coakley, Jeremy Price, Miriam Simun, Andrew Weinberger, Jim Richardson, and Lars Ludwig. See http://www.hyperorg.com/blogger/2010/08/12/2b2k-suggestions-wanted-amateur-scientists/#comments.

24. See http://www.galaxyzoo.org/story.

25. See http://watch.birds.cornell.edu/nestcams/clicker/clicker/index.

26. See http://www.networkedorganisms.com.

27. Luigi Folco et al., "The Kamil Crater in Egypt," *Science,* June 29, 2010, http://www.sciencemag.org/cgi/content/abstract/sci;science.1190990v1.

28. Jess McNally, "Asteroid Crater Hunting from Your Home," *Wired Science,* August 10, 2010, http://www.wired.com/wiredscience/2010/08/crater-hunting/. See also "The Physics arXiv Blog" post for August 10, 2010, at the *MIT Technology Review,* http://www.technologyreview.com/blog/arxiv/25583/.

29. Sarah Reed, "Astronomical Find by Three Average Joes," *Science,* August 12, 2010, http://news.sciencemag.org/sciencenow/2010/08/astronomical-find-by-three-avera.html.

30. Seth Cooper, Firas Khatib, and Adrien Treuille et al., "Predicting Protein Structures with a Multiplayer Online Game," letters section of *Nature* 466 (August 5, 2010): 756–760, http://www.nature.com/nature/journal/v466/n7307/full/nature09304.html.

31. See http://www.patientslikeme.com/about. Also, http://www.als.net/. Also, Deborah Halber, "Stephen Heywood, Son of Prof. John Heywood, Dies at 37," *MIT News,* November 28, 2006, http://web.mit.edu/newsoffice/2006/obit-heywood.html.

32. Hawkins's dissertation was on recognizing handwriting, so he had the satisfaction of seeing that work successfully commercialized in the Palm Pilot that he invented.

33. Interview with the author, October 25, 2010.

34. Alexia Tsotsis, "Attempt at P ≠ NP Proof Gets Torn Apart Online," August 12, 2010, TechCrunch, http://techcrunch.com/2010/08/12/fuzzy-math/. See also Vinay Deolalikar's blog at http://www.hpl.hp.com/personal/Vinay_Deolalikar/.

35. Tsotsis, "Attempt at P ≠ NP Proof Gets Torn Apart Online."

36. Richard Smith, "The Power of the Unrelenting Impact Factor—Is It a Force for Good or Harm?" *International Journal of Epidemiology* 35, no. 5: 1129–1120, DOI:10.1093/ije/dyl191, http://ije.oxfordjournals.org/cgi/content/full/35/5/1129.

37. Ibid.

38. Interview with Victor Henning, August 18, 2010.

39. The number of active users and nonduplicated articles is, of course, lower.

40. Interview with Peter Binfield, October 8, 2009.

41. Interview with Jean-Claude Bradley, August 25, 2010.

42. The UsefulChem blog is at http://usefulchem.blogspot.com/. There is a wiki at http://usefulchem.wikispaces.com/.

43. Jean-Claude Bradley, "Secrecy in Astronomy and the Open Science Ratchet," July 11, 2010, http://usefulchem.blogspot.com/2010/07/secrecy-in-astronomy-and-open-science.html; Alan Boyle, *The Case for Pluto* (John Wiley & Sons, 2010).

44. Interview with Peter Binfield, October 8, 2009.

45. Email from Peter Suber, February, 24, 2011.

46. Bruno Latour, *Pandora's Hope* (Harvard University Press, 1999).

47. Popper's *Logik der Forschung* was published in German in 1934. He translated it into English as *The Logic of Scientific Discovery* in 1959.

48. Michel Foucault, *The Archaeology of Knowledge,* translated by Alan Sheridan (Psychology Press, 2007). See also Linda Martin Alcoff, "Foucault's Philosophy of Science: Structures of Truth/Structures of Power," in Gary Gutting, ed., *Continental Philosophy of Science* (Wiley-Blackwell, 2005), pp. 211–223, http://www.alcoff.com/content/foucphi.html.

49. See Juan Carlos Castilla-Rubio and Simon Willis, "Planetary Skin: A Global Platform for a New Era of Collaboration," March 2009, http://www.cisco.com/web/about/ac79/docs/pov/Planetary_Skin_POV_vFINAL_spw_jc_2.pdf.

50. Interview with Juan Carlos Castilla-Rubio, March 17, 2010.

51. Interview with Timo Hannay, February 12, 2010.

52. Interview with John Wilbanks, December 14, 2009.

53. Richard Rorty, *Philosophy and the Mirror of Nature* (Princeton University Press, 2009).

54. Richard Rorty, *Consequences of Pragmatism* (University of Minnesota Press, 1986), p. xvii.

55. "Climate of Fear," *Nature* 464, no. 141 (March 11, 2010), DOI:10.1038/464141a, http://www.nature.com/nature/journal/v464/n7286/full/464141a.html.

56. See Mary Elizabeth Williams, "Jenny McCarthy's Autism Fight Grows More Misguided," *Salon*, January 6, 2011, http://www.salon.com/life/feature/2011/01/06/jenny_mccarthy_autism_debate.

57. Steven Pinker, "Mind over Mass Media," *New York Times*, June 10, 2010, http://www.nytimes.com/2010/06/11/opinion/11Pinker.html. For a rebuttal, see Nicholas Carr's blog post, "Steven Pinker and the Internet," June 12, 2010, http://www.roughtype.com/archives/2010/06/steven_pinker_a.php.

58. David Quammen, *The Reluctant Mr. Darwin: An Intimate Portrait of Charles Darwin and the Making of His Theory of Evolution* (W. W. Norton, 2007), p. 84.

59. Ibid., p. 76.

60. Ibid., p. 141.

61. Ibid., p. 137.

62. Ibid., p. 162.

63. Ibid., p. 172.

64. Jean-Claude Bradley, "Dangerous Data: Lessons from My Cheminfo Retrieval Class," January 2, 2010, http://usefulchem.blogspot.com/2010/01/dangerous-data-lessons-from-my-cheminfo.html.

65. "Eggs Good for You This Week," *The Onion*, April 28, 1999, http://www.theonion.com/articles/eggs-good-for-you-this-week,4144/.

Chapter 8: Where the Rubber Hits the Node

1. Jack Welch, with John A. Byrne, *Jack: Straight from the Gut* (Warner Business Books, 2001) p. 103.

2. "Fortune Selects Henry Ford Businessman of the Century," November 1, 1999.

3. Welch, *Jack: Straight from the Gut*, p. 101.

4. Ibid., pp. 100–103.

5. Ibid., p. 103.

6. Interview with Rob Stanton, February 22, 2010.

7. I wrote about West Point and leadership on the *Harvard Business Review* blog: "'Powering Down' Leadership in the U.S. Army," November 2010, http://www.eiu.com/index.asp?layout=EBArticleVW3&article_id=1057614890. See also Tony Burgess's blog post on that site: "What's Your Primary Focus: Leadership or Effectiveness?" May 11, 2010, http://blogs.hbr.org/imagining-the-future-of-leadership/2010/05/whats-your-primary-focus-leade.html.

8. From the Wikipedia entry "Virginia Tech Massacre," http://en.wikipedia.org/w/index.php?title=Virginia_Tech_massacre&oldid=418336016.

9. Andrea Forte and Amy Bruckman, "Scaling Consensus: Increasing Decentralization in Wikipedia Governance," *Proceedings of HICSS [Hawaii International Conference on System Sciences]*, Waikoloa, Hawaii, January 2008, http://dlc.dlib.indiana.edu/dlc/bitstream/handle/10535/5638/ForteBruckman ScalingConsensus.pdf%3Fsequence%3D1.

10. "Virginia Tech Massacre," http://en.wikipedia.org/w/index.php?title= Virginia_Tech_massacre&oldid=418336016.

11. See http://en.wikipedia.org/wiki/Wikipedia:NOT.

12. Quoted in Forte and Bruckman, "Scaling Consensus," p. 6.

13. See http://en.wikipedia.org/w/index.php?title=Wikipedia:Notability& oldid=386469793.

14. Interview with Jimmy Wales, October 1, 2010.

15. Ibid.

16. Kristie Lu Stout, "Reclusive Linux Founder Opens Up," World Business section of CNN.com, May 29, 2006, http://edition.cnn.com/2006/BUSINESS /05/18/global.office.linustorvalds/. One guess at the number of developers who worked on Windows 7 is 1,000, based on a post by Steve Sinofsky at the "Engineering Windows 7" blog on August 17, 2008 (http://blogs.msdn.com/b/e7 /archive/2008/08/18/windows_5f00_7_5f00_team.aspx). But it is extremely difficult to do an apples-to-apples comparison, so to speak.

17. Eric Steven Raymond, "The Cathedral and the Bazaar," http://catb.org/ ~esr/writings/homesteading (May 1997). Raymond later published a book by the same name.

18. "The Scalability of Linus," Slashdot, July 23, 2010, http://linux.slashdot .org/story/10/07/23/123209/The-Scalability-of-Linus?

19. See http://www.debian.org/devel/constitution as well as Mathieu O'Neil, *Cyberchiefs: Autonomy and Authority in Online Tribes* (Pluto Press, 2009), p. 132.

20. O'Neil, *Cyberchiefs*.

21. Interview with Noel Dickover, February 25, 2011.

22. See http://www.fantasypumpkins.com/.

23. Peter J. Denning and Rick Hayes-Roth, "Decision Making in Very Large Networks," *Communications of the ACM [Association for Computing Machinery]* 49, no. 11 (November 2006): 19–23.

24. There are important exceptions. For example, Charlene Li describes how Cisco's CEO, John Chambers, has driven decision-making down several layers. See Charlene Li, *Open Leadership: How Social Media Can Change the Way You Lead* (Jossey-Bass, 2010), pp. 40–43, 250–252.

Chapter 9: Building the New Infrastructure of Knowledge

1. Michael Barker, at the Harvard University Library, confirmed this as a ballpark figure in an email dated March 3, 2011.

2. James Crawford, "On the Future of Books," October 14, 2010, *Inside Google Books* blog, http://booksearch.blogspot.com/2010/10/on-future-of-books.html.

3. Robert Darnton, "A Library Without Walls," *New York Review of Books*, October 10, 2010, http://www.nybooks.com/blogs/nyrblog/2010/oct/04/library -without-walls/. Disclosure: I am a member of the Digital Public Library of America's "technical workstream," and the library lab that I co-direct will have entered the DPLA's call for project ideas before this book is printed.

4. Kevin Kelly, *What Technology Wants* (Penguin, 2010).

5. James Aitken Wylie, *The History of Protestantism with Five Hundred and Fifty Illustrations by the Best Artist*, Vol. 1 (Cassell, 1899), p. 113, http://books .google.com/books?id=kFU-AAAAYAAJ.

6. See Ethan Zuckerman's excellent post "Shortcuts in the Social Graph," October 14, 2010, http://www.ethanzuckerman.com/blog/2010/10/14/shortcuts-in -the-social-graph/.

7. During the 2008 presidential campaign, Sarah Palin was accused of pressuring a local librarian to censor some books. See Rindi White, "Palin Pressured Wasilla Librarian," *Anchorage Daily News*, September 4, 2008, http://www.adn .com/2008/09/03/515512/palin-pressured-wasilla-librarian.html.

8. Tim Berners-Lee, "Linked Data," July 27, 2006, http://www.w3.org/Design /Issues/LinkedData.html.

9. This was the price quoted at Fisher Scientific on June 11, 2011. See http:// www.fishersci.com/ecomm/servlet/itemdetail?catalogId=29104&productId= 3426224&distype=0&highlightProductsItemsFlag=Y&fromSearch=1&store Id=10652&langId=-1.

10. See http://www.dublincore.org.

11. Jillian C. York, "The False Poles of Digital and Traditional Activism," September 27, 2010, http://jilliancyork.com/2010/09/27/the-false-poles-of-digital -and-traditional-activism/.

12. See Mathieu O'Neil's *Cyberchiefs: Autonomy and Authority in Online Tribes* (Pluto Press, 2009) for a discussion of authority and the new online institutions. See also danah boyd's talk, "The Not-So-Hidden Politics of Class Online," at Personal Democracy Forum, New York City, June 30, 2009, http://www .danah.org/papers/talks/PDF2009.html.

13. Serge Schemann, "Clamor in the East; East Germany Opens Frontier to the East for Migration or Travel; Thousands Cross," November 10, 1989, http:// www.nytimes.com/1989/11/10/world/clamor-east-east-germany-opens-frontier -west-for-migration-travel-thousands.html?scp=4&sq=berlin+wall&st=nyt.

14. E. J. Dionne, "The Political Campaign: From Politics Barely a Pause: Candidates Already Living '88," *New York Times,* Nov. 10, 1986, http://www .nytimes.com/1986/11/10/us/the-political-campaign-from-politics-barely-a -pause-candidates-already-living-88.html?scp=1&sq=&st=nyt.

15. From the abstract on the *New York Times* site: http://select.nytimes.com /gst/abstract.html?res=F5061EFA3454127A93C2A8178AD95F4D8285F9&scp =7&sq=&st=p.

16. See the Open Courseware Consortium site: http://www.ocwconsortium .org/.

17. E. Hargittai, "The Digital Reproduction of Inequality," in David Grusky, ed., *Social Stratification* (Westview Press, 2008), pp. 936–944. See also E. Hargittai, "Digital Na(t)ives? Variation in Internet Skills and Uses Among Members of the 'Net Generation,'" *Sociological Inquiry* 80, no. 1 (2010): 92–113.

18. Interview with Howard Rheingold, August 2, 2010. See his forthcoming book, *Net Smarts: How to Thrive Online* (MIT Press, 2012).

19. Dan Gillmor, *Mediactive* (2010), http://www.mediactive.com.

20. See Ethan Zuckerman's book, untitled at this writing.

21. For the wisest and biggest-hearted thinking on this topic, follow Ethan Zuckerman's writings, including his upcoming book as well as his blog at http://www.ethanzuckerman.com/blog/.

22. Clay Shirky, *Cognitive Surplus* (Penguin Press, 2010).

23. Robert Morris pointed me to this field.

24. Richard G. Pearson, "Species Distribution Modeling for Conservation Educators and Practitioners" (2007), available at biodiversityinformatics.amnh .org/ . . . /SpeciesDistModelingSYN_1–16–08.pdf, pp. 40–41. See also C. J. Raxworthy, E. Martinez-Meyer, N. Horning, R. A. Nussbaum, G. E. Schneider, M. A. Ortega-Huerta, and A. T. Peterson, "Predicting Distributions of Known and Unknown Reptile Species in Madagascar," *Nature* 426 (December 2003): 837–841.

The bibliography of this book can be found at http://www.TooBigToKnow.com.

INDEX